TELEMEDICINE TECHNOLOGIES

TELEMEDICINE TECHNOLOGIES

INFORMATION TECHNOLOGIES IN MEDICINE AND TELEHEALTH

Bernard Fong
Centre for Prognostics and System Health Management, City University of Hong Kong

A.C.M. Fong
Auckland University of Technology, New Zealand

C.K. Li
Hong Kong Polytechnic University and Centre for Prognostics and System Health Management, City University of Hong Kong

A John Wiley and Sons, Ltd., Publication

Registered office
John Wiley & Sons Ltd, The Atrium, Southern Gate, Chichester, West Sussex, PO19 8SQ, United Kingdom

For details of our global editorial offices, for customer services and for information about how to apply for permission to reuse the copyright material in this book please see our website at www.wiley.com.

Library of Congress Cataloging-in-Publication Data

Fong, Bernard.
 Telemedicine technologies : information technologies in medicine and telehealth / Bernard Fong, A.C.M. Fong, C.K. Li.
   p. ; cm.
 Includes bibliographical references and index.
 ISBN 978-0-470-74569-4 (cloth)
 1. Telecommunication in medicine.   I. Fong, A.C.M.   II. Li, C. K. (Chi Kwong), 1952- III. Title.
 [DNLM: 1. Telemedicine.   2. Medical Informatics Applications. W 83.1 F674t 2010]
 R119.9.F66 2010
 610.285–dc22

                                                                                    2010023379

A catalogue record for this book is available from the British Library.

Print ISBN: 9780470745694 (hb)
ePDF ISBN: 9780470972144
oBook ISBN: 9780470972151

Typeset in 10/12pt Times by Aptara Inc., New Delhi, India

Contents

List of Figures

Foreword

Over the past couple of decades, advancements in information and communication technologies have brought medical services to virtually all corners of the world. For example, a surgeon can now carry out a surgical operation outside the operating theatre, and a physiotherapist can monitor the progress of post-surgical rehabilitation without visiting the patient. Technologies do not only assist medical practitioners and patients receiving treatment, they also benefit perfectly healthy people by providing a wide range of general health assessments. This can help maintain optimum health and identify abnormalities as early as possible via prognostics and health management techniques.

Written by three experts in the areas of telemedicine, multimedia and knowledge management, this book comprehensively covers aspects of telemedicine applications from emergency rescue to treatment and health monitoring and advanced disease detection. The text provides readers with fundamental knowledge in data communications without extensive mathematics, followed by a number of application areas.

This book is primarily intended for readers ranging from medical professionals to final year undergraduate and first year graduate students in biomedical engineering or related disciplines. One of the book's main objectives is to help medical practitioners acquire fundamental knowledge in the technology behind the systems that help them with their work, and to serve as a reference for people who design and implement telemedicine systems. The text provides a detailed coverage of how technological advancements in high-speed wireless networking for secure transmission of medical information may benefit both healthcare professionals and end users, ranging from telecommunication technologies from small body area networks to home and global enterprise networks between entities such as hospitals and different authorities and agencies.

M. G. Pecht
Visiting Professor and Director, Centre for Prognostics and
System Health Management, City University of Hong Kong, and
Professor and Director, CALCE, University of Maryland, College Park, MD, USA

Preface

Telemedicine is the broad description of providing medical and healthcare services by means of telecommunications. Information Technology (IT) in areas covering control, multimedia, pattern recognition, knowledge management, image and signal processing; have enabled a wide range of applications to be supported.

The combined effect of worldwide population growth and aging population in most developed nations gives rise to the soaring demand on the public health system. The impact on national health systems in many countries is further fueled by change in lifestyle and environmental pollution. All these are stretching health systems to their limits. This is evident from the trend of chronic disease and obesity-related complications affecting younger people over the past decade. The economic prosperity now enjoyed by many is a direct result of hard work by the previous generation and excessive consumption of natural resources that may bring a range of problems to future generations. In response to all these, we take good care of the senior citizens who have devoted decades of their lives to ensure today's prosperity. On the same token, we are working hard to enhance medical technologies to improve our health, and to provide a sustainable healthcare system for the next generation. Telemedicine is one of the key solutions to fulfilling our responsibilities for the young and the elderly alike.

There is an emergent interest amongst government authorities, healthcare service providers, academia, medical devices and supplies industries to optimize the efficiency of providing a wide range of medical services in terms of both cost and time. The effective utilization of telemedicine and related technologies will be able to assist with, but is not limited to:

- support more types of services
- bring services to more people in more regions
- make healthcare more affordable for the poor and the elderly
- optimize health for all ages
- on-scene treatment for medical professionals on the move
- provide preventive care in addition to emergency treatment
- remote rehabilitation monitoring
- chronic disease relief and care
- ascertain service reliability and eliminate human errors
- safeguard patients' information and medical history.

To address the growing trend of telemedicine deployment in both urban and rural areas throughout the world, this book discusses different technologies and applications surrounding telemedicine and the challenges faced. This book also looks at how various signs of a human body are captured and subsequently processed so that they can be well used for providing treatment and health monitoring. As conventional medical science tends to provide remedies according to symptoms, we also explore how technologies in alternative medicine can go down to the fundamentals to address the root cause by optimizing health in general.

Book Overview

Chapter 1 is an introductory chapter that provides a general picture of what telemedicine entails and the importance of providing quality healthcare in various areas of medical practicing with the aid of telemedicine technologies. The underlying concepts in various areas are concisely discussed and most of these will be elaborated in more depth throughout the book. The technologies associated with individual application would depend on current technology availability and specific regulatory limitations imposed by respective authorities of a given country. Readers should be able to get a good understanding about how medical and IT professionals are linked closely together through technological advancements. It is about how they help each other work better. And more importantly, how the general public improves the way they enjoy better health and medical services as a result of technology.

Chapter 2 provides technical coverage on what telecommunication technology is all about, and how it can be applied to better healthcare. Although this chapter primarily provides technical knowledge to readers, we shall not go deep into the engineering and mathematical aspects as the main scope on this book is technologies related to medical and healthcare applications. However, adequate knowledge will be provided to make use of underlying communications technology for healthcare. We will see what solutions are currently available and how to select the type of network most suitable for a given telemedicine application. Examples will be given to demonstrate how technology is applied. We will also look at the harsh outdoor environment where wireless communication systems will be affected by various factors. Fundamental limitations of technology will be dealt with so that what can be done or cannot be done will also be discussed.

Chapter 3 first looks at how life saving can be accomplished with technology developed for emergency rescue. We then look at wireless communication systems used in remote patient monitoring. This is a particularly important application for servicing rural areas and the elderly. Such technology is also suitable for rehabilitation so that patients can recover at home with the assurance that they are properly looked after even after they are discharged from the hospital. Various topics on body area network will be considered. These include different types of wearable monitoring devices, body sensors, data communication between devices, and practical difficulties faced.

Chapter 4 discusses the information theory behind successful representation of various types of medical information with binary bits. We start by looking at different ways of collecting data from patients; different applications would require very different types of capturing devices. For example, measuring a person's heart rate and electrocardiograph (ECG) would require very different instruments. We then look at precautions necessary for medical data transmission and storage, followed by storage applications such as electronic patient records and electronic pharmacy.

Chapter 5 considers system deployment issues with an example on wireless telemedicine system development. It deals with a number of possible options and the importance of ensuring quality and reliability, something particularly important in life saving critical missions.

Chapter 6 introduces the concept of information security and how to implement secured telemedicine systems for different applications. Patients' privacy must be respected and any information collected needs to be safeguarded throughout the entire process from collection to analysis and subsequent storage. There have been reported cases of serious misconduct due to medical personnel losing removable storage devices containing patients' information like thumb drives and memory cards. These irresponsible acts can be easily restrained by providing secured remote access to hospital staff. Any data collected for statistical analysis must ensure individual persons cannot be identified so that all such information remains anonymous. Since certain data needs to be shared between medical institutions and government agencies, mechanisms for maintaining data accuracy as well as anonymity is always crucially important. Before leaving this chapter on data security, we will look at the evolvement of technologies related to biometric identification.

Chapter 7 introduces alternative medicine and may not be too relevant to certain regions although it is increasingly accepted as an effective way to treat prolonged illnesses such as colds, coughs and asthma. This chapter therefore aspires to give readers some background information on what alternative medicine entails and how information technology can be applied to serving the community better through practicing alternative medicine. We also look at an example of using biomedical databases for herbal medicine and acupressure aimed at treating patients who may require long term treatment. The discussion will then proceed to technology in optimizing health, like progress monitoring in gymnasiums or just taking a short morning jog. Consumer healthcare products such as foot spa and massage chairs are becoming increasingly popular throughout the world. These products offer many new features including integration with existing audio/visual systems and other home appliances. We will look at how related technologies help improve quality of life and maintain optimal health.

Chapter 8 addresses the issues of providing electronic healthcare from a user's point of view. This is considered to be an important part because as population ageing becomes a more serious problem in most developed countries the demand for these services is expected to grow tremendously over the next few decades. Eventually all of us will become old and require more medical attention during our natural ageing process. Through utilization of technology we will pay fewer visits to clinics and hospitals, and we will be better looked after. People living in rural areas will find this particularly helpful since not all remote small towns have medical facilities readily available at all times. Telecare becomes an important telemedicine application for providing easy access of healthcare to those of special needs. Although technology may not always reduce the risk of accidents occurring, we do have mechanisms for keeping an eye on people who need caring so that necessary actions can be taken without delay should a mishap occurs. In addition to providing special care to the elderly and those with special needs, we also look at how technology can help people recover from sports injury. Some exercise may facilitate speedy recovery yet improper movement can worsen the affected area. So, technology that monitors the rehabilitation progress would therefore be very helpful for those who struggle to recover from injuries.

Chapter 9 begins with an overview on how medical data can be stored and transmitted in a more efficient way by using different coding and compression techniques. We consider other applications such as learning support for medical and nursing students as technologies make training of healthcare professionals easier and more efficient. We also explore other emerging technology for telemedicine advancements such as haptic sensing by conducting various tasks through touch, and what future telemedicine and information technology has to offer.

The book concludes with a brief summary of different types of wireless networks that can support various telemedicine applications, and emerging industrial standards that will likely influence how telemedicine systems and related services will evolve over the next few years.

It has been seen over recent years that the capabilities of telemedicine systems have grown tremendously due to advances in information technologies. As a result of numerous published work in telemedicine and related technologies in this rapidly developing subject, may have missed out some of these in the context of this book.

Acknowledgements

First and foremost, the authors wish to thank all readers for taking the time to learn more about telemedicine technologies. The authors are confident that this book will enlighten readers to develop their expertise in further enhancing medical and healthcare technologies to benefit more people in the community. The main objective of telemedicine technologies has always been to extend medical services to more areas for more people so that people can live healthier and longer irrespective of where they are.

Over the years, the authors have seen numerous cases where people are unable to enjoy accessible healthcare either because they cannot afford it or service cannot be extended to their areas due to a number of reasons. The continuing advances of telemedicine technologies that break the geographical barrier of providing quality healthcare urged us to write a book to share our insights together with the underlying technologies that can potentially benefit millions, if not billions, of people. Much of what we have learned over the years comes as a direct result of taking care of our retired parents as well as our delightful children, all of whom, in their unique ways, inspired us and subconsciously contributed a tremendous amount to the content of this book on promoting enhancement of telemedicine technologies to help people of all ages.

We must also thank Professor Michael Pecht who taught us the importance of addressing reliability issues in the book, how to look at reliability in a different context and the art of applying prognostics and health management techniques to medical systems.

We also have to thank Professor Nirwan Ansari, with whom I have had the great pleasure of working with in wireless networking research. Over the years he has taught us a lot about advanced networking technologies and applying such knowledge to telemedicine systems and services.

We also wish to thank Anna Smart, Sarah Tilley and Tiina Ruonamaa, the editorial team at John Wiley with a profusion of patience and talent whose excellent work has led to a significant improvement on the presentation of the book.

Finally, the authors wish to thank Ocean Blue Software (UK) Ltd for permission to reproduce copyright material of the case study illustrated in Chapter 9. Every effort has been made to trace rights holders. However, in case any have been inadvertently overlooked, the authors would be pleased to make the necessary arrangements at the very earliest opportunity.

1

Introduction

1.1 Information Technology and Healthcare Professionals

The history of modern telemedicine goes back to the invention of the traditional telephone about a century ago. Medical advice was given by physicians over the telephone. The term *telemedicine* is very simply a description of supporting medical services through the use of telecommunications. 'Tele' is a prefix for distant, originated from ancient Greek. So, telemedicine literally translates to providing medical services over distance. Telecommunications used in medical applications can be categorized as sending medical information between a pair of transmitters and receivers. The so-called 'medical information' can be as simple as a doctor providing consultation to sophisticated data captured from a human body. In its most primitive form, 'The Radio Doctor' first appeared in the *Radio News* magazine (circa 1924) and is perhaps the earliest documented case of utilizing telecommunication technology for medical application. Although information technology has been used in healthcare since then, (Moore, 1975) was the first scientific literature formally addressing the application of technology in medicine that appeared.

As information technology advanced over the past decades, a wider range of healthcare services could be supported. Indeed, the types of services that can be supported is so vast that any book which makes an attempt to provide comprehensive coverage of all areas will most likely contain thousands of pages in several volumes. This book aims to provide an in-depth coverage on how wireless communications and related technologies are used in medical services, we will also look at the challenges and limitations of current technology associated with healthcare information systems.

We will first begin by taking a look at how simple wireless communication networks function and what a telemedicine system consists of. We look at a number of examples that describe how a primitive system supports healthcare services. In the course of the book, more sophisticated systems will be described in more detail.

The context of this chapter is anticipated to give readers an overview on how information technology is widely used in assisting healthcare without going into technical depth. To begin our discussion, we revisit the term 'information technology', something often associated with computer science. Essentially, it is extensively interpreted as a blend of computing and telecommunications. This leads to the acronym ICT, which stands for Information and

Telemedicine Technologies: Information Technologies in Medicine and Telehealth Bernard Fong, A.C.M. Fong, and C.K. Li
© 2011 John Wiley & Sons, Ltd

Communications Technology, also known as infocomm for short. All these are merely descriptions of the use of technology to securely and reliably transmit information between two or more entities. Information Technology (IT) is widely used in many areas that influence our daily life. For example, banking, transportation, manufacturing, etc. This list is seemingly endless. When we see so many information technologies support so many things that we use on a daily basis, it will not be difficult to understand how widely it can be used to support healthcare and medical applications.

Since the information technology and 'dot-com bubble burst' in 2000, the whole IT industry has never quite picked up. Looking at NASDAQ that peaked at its all time high of 5132 in March 2000, then retreated to about a quarter of its peak some nine years later, it does appear that people related to the IT industry have suffered substantial thwack for many years. IT professionals who developed a career in finance enjoyed a few more years of wealth until the subprime mortgage fallout that started in early 2007. So, despite various aspects of IT being widely used in different aspects of daily life, it has a close relationship with the global economic cycle. In contrast, healthcare and medical service is one of the few domains that have consistent high demand very simply because every single one of us understands the fundamental importance of our own well-being. We know as a matter of fact that without quality health nothing will be important to us. For this simple reason, healthcare naturally becomes an essential part of daily life that will continue to be in high demand for many years in the future.

Having realized the prime importance of healthcare, we go further into how IT is applied to healthcare and medical services. Long before the evolution of information technology, herbal medicine practitioners millennia ago already utilized the most primitive form of information exchange mechanism, namely communication systems to convey messages on medical services. (Wang, 1999) documented a case where Shen Nong made use of information exchange for treatment of respiratory syndrome as far back as 2735 BC, this may not have been the first case, but it is certain that medicine and communications have been linked together for over 4,000 years. As IT becomes more sophisticated over time, a more diverse range of medical services can be supported. To name a few, IT in medicine involves drug prescription, spread of pandemic modelling, patient monitoring, remote operation, medical database and so on. This is by no means an exhaustive list and we will cover these as well as many others throughout the book.

Obviously, healthcare professionals can make use of IT advancements in different areas. Advantages brought by IT include improvement in reliability, efficiency, precision, ease of information retrieval, accomplishing tasks remotely, and better organization. Healthcare therefore becomes more readily accessible and more efficient. We will look at how technology benefits healthcare professionals, with the assumption that readers have very little prior IT knowledge and know virtually nothing about the underlying technologies.

1.2 Providing Healthcare to Patients

In addition to facilitating medical practitioners to perform their tasks, another important issue to address is the healthcare services provided to patients, as they are the end users who must feel comfortable receiving the treatment given. The provision of a technically feasible solution is not the only obstacle to deal with. Other important issues including patients' acceptance and accessibility must also be addressed. We strive to look at providing healthcare solutions to patients using IT from the perspectives of both providers and patients. End users, particularly children and the elderly, may not be too keen on accepting technology as a tool for healing. Convincing patients of the benefits of IT in healthcare may involve liability, security and privacy issues. For example, in the case of monitoring or tracking a patient recovering at home, the patient must be assured that personal information is securely kept and no such information is accessed in any way without consent.

Before leaving the topic on providing care to the elderly for now, it is worth briefly noting the advantages brought to this group of users by telemedicine technology. As population ageing is becoming a more significant concern in many countries, it can be widely expected that more care and monitoring will be needed. A significant increase in the application of wireless communications in elderly care has been seen over the past few years as related technologies become more mature. The cost of service becomes more affordable and portable devices become smaller and more user-friendly. As pervasive computing technology advances, more comprehensive and automated services will become available to the ageing population in the years to come (Stanford, 2002). The design of interconnected devices and sensors on the patients' side must ensure non-obtrusiveness and can be comfortably worn. Also, user's movement will not be restricted and reliability will not be affected irrespective of wearing condition. User-friendliness is another important design factor, as absolutely minimal training should be necessary especially for children and the elderly. These should be genuine 'plug-and-play' devices. In this sense, the healthcare system in the patient's home can be installed by a technician during initial deployment. Thereafter, almost everything should be fully automatic except for unavoidable scheduled maintenance such as battery replacement and calibration.

Let's elaborate more on a patient's point of view as an end user. The primary objective of telemedicine is to provide medical services remotely. Amongst numerous advantages brought to patients by telemedicine, an obvious convenience is reducing the need for clinical visits. Through utilization of IT, a patient can rest at home while receiving full medical attention. Reviewing the level of medical support provided over the past two to three decades, IT has certainly provided tremendous benefits to the general public as a whole. The advancement of faster computers and more efficient bandwidth usage has allowed more types of services to be extended to more users. For example, a few decades ago a simple request for medical advice could be obtained by finding a fixed line telephone and dialing in to the clinic where a physician was stationed. With the availability of mobile Voice over Internet Protocol (VoIP) technology, one can now simply pick up a mobile phone and place a video-enabled call to a physician; the physician does not necessarily have to be situated inside the clinic in order to provide advice. This is just one amongst numerous examples where IT advancements have made healthcare more readily available. More examples will be presented throughout the book.

While the benefits to patients are obvious, there is a wide range of challenges that different parties face in order to serve the patients. These concern people from developers, practitioners, healthcare management and authorities. The subsequent paragraph will highlight challenges that different people face starting from initial planning stage to final rollout and continuing maintenance.

From the IT perspective, the fundamental question is feasibility. Primary consideration is whether current technology is capable of doing something. After this comes practicality and cost effectiveness. We begin by considering an example where school children are to enroll into a program that ensures their school bags are ergonomically prepared to minimize issues with back pain. The advantage to participating children is very obvious because the program should reduce their chances of suffering from back pain. However, how viable is the entire program? We need to understand more about the technology involved in order to answer this seemingly simple question.

In this case study, we have the following parties involved: engineers developing the monitoring system, clinical staff analyzing the captured data, funding bodies providing necessary resources, children participating in the study, and finally, participants' parents giving consent to their children's involvement. We shall look at the case with respect to benefits and concerns from each party's standpoint.

1.2.1   Technical Perspective

Biomedical engineers need to develop a system based on requirements specified by clinical staff, such as that illustrated in Figure 1.1, with the necessary sensors and data communication network. This simple system has a number of sensors forming a sensor network for capturing different types of information about a patient. It is linked to the system for analysis by a workstation and storage in electronic patient record (EPR), and is monitored by necessary system and network administration tools. In this discussion, we shall not go into the technical details whilst giving an insight into what is involved. Engineers analyze this by evaluating technical feasibility and practicality. Digging deeper into technical challenges, one obvious issue to address is how to ensure whatever captured data is meaningful. There are several factors that influence the validity of data, most notably from what the sensors have picked up followed by what has been transmitted and subsequently received. In this respect, the sensors must be securely attached at the relevant points of the participant's body, and each sensor

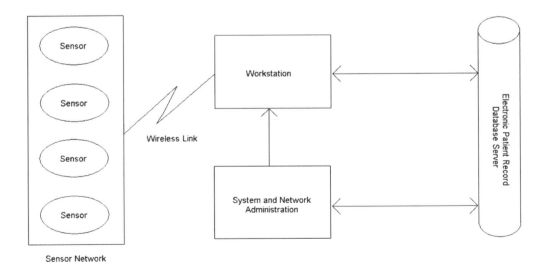

Figure 1.1   A simple biosensor network

must be sensitive enough to pick up any subtle tilting of the body while not too sensitive to pick up any vibration from other sources. Having dealt with these problems, next we must ask whether the sensors are suitable for the specific application, the size may be too large for attaching onto a child, and whether it will cause any discomfort. Are readings affected by any physical obstacles that may be separating the children from the backpack, such as clothing? How is captured data sent out for processing and analysis? Will sensors interfere with each other if placed too close together? Here is just a list of questions related to sensors that need to be dealt with.

We shall proceed by assuming that sensors are well-designed and we manage to overcome all problems listed above. So, we are technically able to capture a set of valid data that tell us something about a child's behaviour when carrying a backpack. We now look briefly at how telemedicine is utilized in a biosensor network; we will come back to this with more details in Section 3.5. In the previous paragraph we raised a question about how the captured data is sent out, essentially we have two choices, namely using wireless communications or connecting the sensors with wires. How they compare will depend on the system itself as there is no clear advantage with either option. This is one major topic that we will cover throughout the book.

Briefly summarizing the discussion here, we have seen how many questions need to be addressed in relation to the deployment of such a supposedly simple health monitoring system. So, although the system may appear simple enough to patients, design and implementation may not be as straightforward and there are so many limitations.

1.2.2   Healthcare Providers

Healthcare professionals should understand that technology is available for making their routine work easier and safer. Many may still prefer traditional practicing methods, just like many people still prefer jotting down notes using pen and paper. Others may find technologies helpful when using a personal digital assistant (PDA) device for the same purpose. There are, of course, many advantages with a PDA although users may need to familiarize themselves with its user interface. Another concern to some is the risk of losing its stored data due to breakdown. We can see that people who are so used to conventional ways of carrying out a task may need to be convinced of the associated benefits technologies bring, in order to impel them into learning to utilize technologies. So, as a practitioner, a simple-to-use interface would be a fundamental design requirement. The entire process should be as automated as possible while maintaining a very high level of reliability. Different applications may have very different demands. For example, tele-surgery requires ultra-high precision for control and crystal clear imaging details with no time delay, whereas tele-consultation may have much less stringent requirements.

Although technical advancements may be more efficient and fault-free enabling numerous tasks to be accomplished quickly and reliably, the incentive of using IT solutions may not be that compelling unless practitioners have mastered the operations of what is made available to them. Getting used to something new, especially for critical tasks, can be a major challenge. A uniform change to new technology for all applications would be vitally important for a swift switch to making good use of available technology.

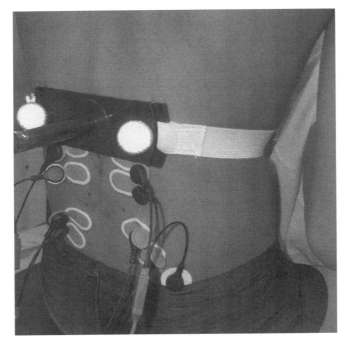

Figure 1.2   Biosensors attached to the back of a patient

1.2.3   End Users

The end users of the system are the patients. The term 'patient' refers to someone who receives medical treatment or service, which includes routine check-up. We should clarify at this point that by definition a person described as a patient may not necessarily be unwell. A perfectly healthy person can be referred to as a patient in this regard. Here, in our case study we have a group of patients who participate in the study of schoolbags on children. They help with the study by having a set of sensors attached to their back while carrying a schoolbag of varying weights. An illustration on how the sensors are attached to the back of a patient is shown in Figure 1.2. We discuss the case study from a patient's point of view by first looking at Figure 1.2. As shown, a number of sensors are attached to the back; each sensor is connected to a data capturing device by a wire. Movement is somewhat affected by the wires so we can readily see the advantages of using wireless sensors as far as the patient is concerned. So, why not wireless? This example exhibits three major technical challenges that make wires extremely difficult to eliminate. First, sensors attached to a child's back must be very small. Powering the sensors can be an issue as installing an internal battery may be a problem. Also, wave propagation issues effectively rule out its use between the body and the bag, as absorption would be a very significant issue. Finally, measurement accuracy given the physical separation of individual sensors and the amount of movement would make the use of wireless solution impractical. For all these reasons, patients have to bear with the wires surrounding them while participating in the experiment.

1.2.4 Authorities

Funding agencies and authorities are most concerned about cost effectiveness. Long term benefits to the community must be clear. In this particular case study, obtaining funding may be difficult despite all the benefits stated in the above sub-sections. This is primarily due to the projected time length of realizing the benefits; this will only be seen when a clear statistical trend of reduction in back pain is attained. The political details are far beyond the scope of the book so we will not discuss anything in detail here. As a general rule, acquiring funding for projects on applying technology to healthcare services, by and large, needs to prove that the benefits will be immediately realized. Further, all these explain a widely seen problem of lacking financial support for rolling out innovative healthcare solutions using technology.

1.3   Healthcare Informatics Developments

In this section, we look briefly at how healthcare and bioinformatics have evolved over the past decades. Medical science has undergone consistent advancements for thousands of years and IT is certainly a much newer topic that has only really commenced from the first computer by Konrad Zuse (circa 1936). Soon after the birth of computers, information storage devices were also born. Health informatics was only made possible when computers were connected together to form a network after computer networking began. The whole idea of health informatics kicked off after World War II as technology became more readily accessible. All these provide a framework to link hospitals together in the cyber world. More recently, computational intelligence makes a wide range of services available. Together with multimedia technology, health and information technology make life-saving and maintaining health something easily accomplished. As illustrated in Figure 1.3, a diverse range of medical and healthcare services can be supported by technology.

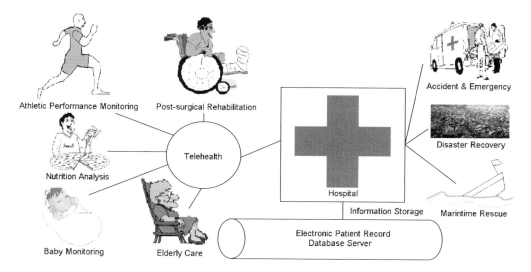

Figure 1.3   Telemedicine supports a range of applications

So, the eight decades after The Radio Doctor appeared have seen the blend of technology with medicine in just about all areas of practice. We have very briefly covered how health informatics has evolved from the first computer and shall pay more attention to more recent developments that are directly related to possible future developments. The first challenges that many people would talk about are probably security and privacy. There are cases of patients' information leaking due to a wide range of reasons from breach of security to loss of storage devices. A significant part of health informatics involves ensuring the security of data keeping, which includes protection from stealing or altering of information and policies ensuring that data will not be misused by parties authorized to access patients' records. A thorough discussion on security and privacy will be presented in Chapter 6. In addition to assurance for safeguarding medical data and privacy, there are many other issues to address since health informatics entails a very wide range of topics in linking people, resources, and devices together and many of these are developed independently over time. The first documented case of modern healthcare informatics deployment in the USA was around the 1950s in a dental project pioneered by Robert Ledley for the National Bureau of Standards (now the National Institute of Standards and Technology) (Ledley, 1965). More medical information systems were developed over the next few years across the USA and most projects were advanced independently of each other. It was therefore practically impossible to develop standards for health informatics systems. The International Medical Informatics Association (IMIA) was formed in 1967 with the main objective to co-ordinate the development of health informatics and related technological advancements. Soon after its formation came the programming language MUMPS (Massachusetts General Hospital Utility Multi-Programming System) for building healthcare applications which is still used today in electronic health record systems. There was soon a need for different variants of the programming languages to run on different computer platforms and a standard was inaugurated in 1974. It is now developed as 'Caché' for medical application development on different computer platforms. It is worth noting that although Caché is still currently used today, many present electronic health record systems are developed using relational databases.

So, a quick look at the development of healthcare informatics reveals that a vast collection of topics in IT are involved. It deals with all aspects of technologies related to preventive caring, consultation, treatment, rehabilitation and monitoring. From this point onwards, we shall concentrate our discussion on communications and networking technologies for healthcare. Related technologies will also be covered from time to time as appropriate.

1.4   Different Definitions of Telemedicine

Telemedicine, the combination of information and communication technologies (ICT), multi-media, and computer networking technologies to deliver and support a wide range of medicine applications and services, has several widely-accepted definitions. The definition given in *wiki* is: 'Telemedicine is a rapidly developing application of clinical medicine where medical information is transferred through the phone or the Internet and sometimes other networks for the purpose of consulting, and sometimes remote medical procedures or examinations.' This definition is simply a brief recapitulation of what is described in Section 1.5 below. Other definitions also exist. For example, the Telemedicine Information Exchange (Brown, 1996) gives its own definition as 'the use of electronic signals to transfer medical data from one site to another via the Internet, telephones, PCs, satellites, or videoconferencing equipment in order to improve access to health care'; and (Reid, 1996) defines telemedicine as 'the use of advanced telecommunications technologies to exchange health information and provide health care services across geographic, time, social, and cultural barriers.'

Variations of definitions do not stop here, the Telemedicine Report to Congress (Kantor, 1997) gives:

'[T]elemedicine can mean access to health care where little had been available before. In emergency cases, this access can mean the difference between life and death. In particular, in those cases where fast medical response time and specialty care are needed, telemedicine availability can be critical. For example, a specialist at a North Carolina University Hospital was able to diagnose a rural patient's hairline spinal fracture at a distance, using telemedicine video imaging. The patient's life was saved because treatment was done on-site without physically transporting the patient to the specialist who was located a great distance away.'

Among these variations of definitions, there are several points in common. First, these are all given in the mid-1990s, suggesting that telemedicine became an important area for just over a decade. Also, all these are closely related to providing different kinds of medical services over distance by utilizing some kind of telecommunication technology.

1.5   Overview on Telemedicine

We have mentioned what telemedicine is at the beginning of the book. Very briefly, it is about the use of telecommunications and networking technologies for transmission of information related to medical and healthcare application. In modern telecommunications information can be transmitted across many types of networks in a variety of forms. By definition, telemedicine can be as simple as two doctors talking about a patient through the telephone or as complex as a sophisticated global hospital enterprise network that supports real-time remote surgical operations with surgeons situated in different parts of the world controlling an operation that takes place in one hospital simultaneously. To elaborate on the vast coverage of telemedicine, Figure 1.4 summarizes a number of services that telemedicine is capable of supporting. It is not a complete list of all services that telemedicine is capable of supporting, but shows all major services currently used worldwide. As we begin looking at these services, it is not difficult to see that there is one thing in common: conveying medical information from one entity to another. Before we proceed further, remember this is an introductory chapter so do not worry about the technical terms and details, as we shall cover them thoroughly throughout the book. Obviously, each application entails different types of information. We look at each of these examples and see what telemedicine does. A simple application like tele-consultation involves delivery of advice, often verbally from an expert to people in need of medical information. In recent years this can extend to services using mobile devices. Tele-diagnosis lets experts carry out diagnostics with medical instruments from a remote location, quite simply by providing a communication link between the two locations. Telemedicine can be far more complex than this, like a sophisticated tele-A&E (Accident and emergency) service which may involve high resolution digital images along with vital signs of a patient collected in a remote location who must be transferred to the hospital with maximum reliability and minimum delay. Some systems may provide additional features such as video conferencing functions and real-time retrieval of medical history records. Likewise, tele-monitoring facilitates monitoring of patients recovering at home or moving around in locations away from the hospital by transmitting different types of data. Depending on the specific application, remote patient

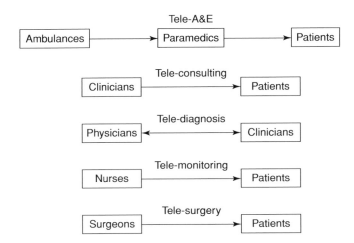

Figure 1.4   Subsets of telemedicine connecting different people and entities together

monitoring may involve the attachment of small wireless biosensors on a patient forming a body area network (BAN) where data captured by an individual sensor is collected within the BAN before being sent out collectively for subsequent processing. In this kind of situation, a telemedicine system may include different types of communication networks. While we shall cover networking in more depth in Chapter 2 with specific emphasis in the following telemedicine applications in Section 2.4, we refer to the example in Figure 1.5 to get some understanding about how three separate networks are interconnected to form a telemedicine system. Here, the patient under observation is surrounded by a BAN which the patient carries when moving around. The data captured is sent to a nearby local area network (LAN) that stores and processes the data. The LAN effectively serves as a bridge between the hospital that is served by the metropolitan area network (MAN) and the patient's home. The LAN is very simply an ordinary home network that is permanently installed at the patient's home. Through installation of appropriate equipment associated with the BAN and establishing a connection to the hospital via the MAN, a telemedicine system that performs tele-monitoring can be set up.

Tele-surgery is probably the most convoluted application partially because of the precision involved. In order to perform a surgical operation from a remote location, the apparatus must have a very high degree of movement in all directions and an unobstructed view must be delivered to the surgeon with good clarity. Therefore, the following basic requirements must be fulfilled to perform even a simple operation:

- Sensors capable of capturing slight movement of a surgeon's hand in real time with extreme precision.
- Cameras that can deliver crystal sharp images of the patient without any obstruction, this is particularly challenging as movement of surgical tools must be taken into consideration, maintaining a good view of the patient at all times is vitally important.
- Actuators that exactly replicate 3-D hand movements as interpreted by the sensors with no time delay.

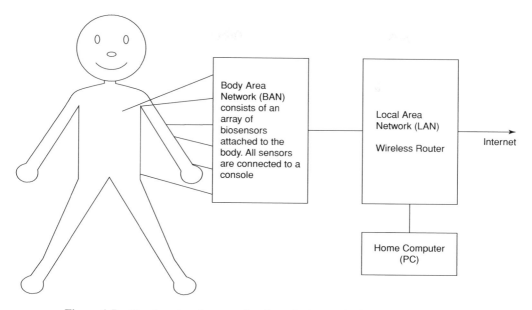

Figure 1.5   Simple network connection from the human body to the outside world

- A communication network that is fast enough to deliver all types of data in both directions, and reliable enough to ensure that it is free of transmission errors throughout the entire operation

By now we should be convinced that telemedicine entails technologies far more than simply POTS (plain old telephone system) that allow two medical professionals to share information verbally. In the later chapters we shall look at more telemedicine applications and underlying technologies that make telemedicine possible.

Connecting people and resources together for better healthcare covers more than the examples given above. We have described ways that the general public can directly benefit from telemedicine; there are other applications such as connecting relevant authorities worldwide to track the spread of diseases in epidemiological surveillance that had been found to be effective in limiting the crisis caused by severe acute respiratory symptom (SARS) and avian influenza (bird flu) over the past few years. Another less obvious yet important application in safeguarding the community is tele-psychiatry where psychiatrists are able to monitor acutely anxious patients so as to proactively prevent violent crimes using telemedicine.

Telemedicine covers almost all aspects of daily life. For example, we can easily access healthcare information by the touch of a 3G cellular phone; getting nutrition information for a healthy diet while dining out has never been easier. Throughout the book we will see telemedicine virtually support all aspects of healthcare in daily life for consumers with a portable device such as cellular phone or notebook computer.

1.6   The Growth of the Internet: Information Flooding in E-Health

We all know what the Internet is, and almost certainly we access the Internet on a daily basis. It is widely perceived that the Internet allows e-mail access, video conferencing, information retrieval from websites, contents downloading of music, video clips, pictures, etc. The evolution of the Internet provides information sharing with worldwide coverage. In essence, long sequences of binary bits '1's and '0's are carried across the world, trillions of them per second. Although only two possible states are sent in the digital world, combinations of these can represent virtually anything that one can imagine. Internet is about integration of devices and information together. In the cyber world information can travel across any part of the world in a fraction of a second. To get a better understanding of how advances of Internet technology support telemedicine we first look at the Internet's development from its birth and what it offers telemedicine.

The origin of Internet was likely the Galactic Network documented by (Licklider, 1962). We can see that telemedicine has a far longer history than the Internet yet the impact of Internet growth on telemedicine advances is very significant. This forms the basis of connecting computers and devices together. Along with the development of packet switching, (Kleinrock, 1961) eventually evolves to networks capable of carrying different types of data to be delivered across a single transmission medium. With such capability, communication networks can support telemedicine in many areas, such as:

- Reliability: quality of service (QoS) assurance.
- Information Sharing: medical web pages online.
- Audio: tele-consultation, respiratory, cardiac and pulmonary sounds.
- Still Images: X-ray, scans, medical images.
- Video Images: tele-conferencing, tele-psychiatry, medical education.
- Databases: electronics patient records, e-pharmacy, alternative medicine.
- Vital Signs: ECG, EEG analysis and storage.

The Internet, in its early days, supported primitive services such as BBS (Bulletin Board System) and e-mail. These were fairly adequate for tele-consultation services. It was not until 1984 when the Internet incorporated TCP/IP (Transmission Control Protocol and the Internet Protocol) that multimedia data traffic was supported.

Since the beginning of the 'modern' Internet that supports all types of telemedicine services described above, there are still threats to the development of telemedicine that exist today. Interestingly, a computer virus that spreads across the Internet may in certain aspects replicate epidemiological control that we have already briefly mentioned. A computer virus is defined as a program that interferes with a computer's normal operation if infected. There are many ways that viruses can spread across the Internet and very commonly they are transmitted as e-mail attachments. Viruses can be disguised in various forms as embedded in other programs or files such as pictures and video clips. They can also be concealed in illicit software just like a human carrying the hepatitis virus and who looks just as a healthy person from the outside. It is well-known that anti-virus utilities can be installed on a computer to safeguard it from virus attacks, telemedicine can actually do something similar in preventing bacterial and viral infections from spreading by proactively tracking down the pattern of spread as well as the mutation of viruses using signal processing techniques.

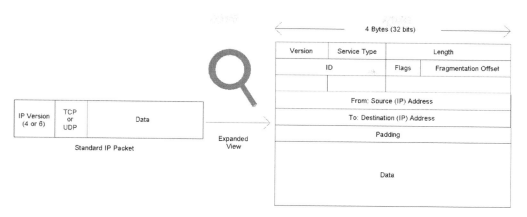

Figure 1.6   Simplified structure of a typical data packet

Development of wireless communications technology allows more flexible deployment of telemedicine for supporting off-site applications. More recently with the advancements of related technologies such as batteries and antennas wearable devices have been made widely available for many medical and healthcare applications. These open up numerous opportunities to new telemedicine services as data can reach almost anywhere while on the move.

Now that we have seen how many different telemedicine application types are espoused the Internet, what is really needed for supporting telemedicine? The Internet does appear to support an unlimited amount of data flowing to virtually anywhere in the world. Of course this perception is not exactly true. The Internet will become saturated when too much data is dumped into it. For a start, telemedicine is about healthcare across the entire world. It does not necessarily mean all medical knowledge should be made available there. Flooding the networks with information will cause it to slow down and malfunction, eventually causing data loss. The Internet must be used in a responsible way since it is a shared medium, minimizing overheads is therefore an important task for telemedicine system developers. Determining what kind of information should be sent requires an understanding of data composition. Data is sent across the Internet as *packets*, a packet is a unit of binary bits sent from the source to the destination. Figure 1.6 illustrates the simplified structure of a typical data packet that is sent across the Internet (Mullins, 2001). It shows that only a portion of the packet contains the actual information that needs to be delivered. The remaining bits are *overheads* that facilitate the transmission of information. Very similar to sending a letter through the postal system, we put the piece of paper that contains our actual message into an envelope, and the envelope contains things like Sender's Address (source location), Destination Address (recipient location), Airmail Label (delivery method), and Postage Stamp (class of service). The pair of flags replicates the envelope itself indicating the packet's enclosure, the protocol defines the delivery method, and the type of service marks the class of service. Finally, we also have the source and destination addresses and, of course, the actual information. In addition, checksum is used for checking data integrity upon receipt, and additional services similar to registered or courier post are also available in the digital networking world. Certain communication protocols provide guarantees for successful delivery, and different QoS schemes can be set to prioritize data traffic across the network.

We see that a data packet contains far more than the actual information. However, we need to bear in mind that we cannot change the way data is structured as it is necessary to comply with applicable standards for data transmitting across the Internet (currently IPv4, and IPv6 is becoming available). What we need to do is to ensure that telemedicine services, especially when utilizing the Internet, should incur minimal overhead. We shall revisit the topic of transmission efficiency shortly in the next chapter. In summary, what we have seen in this section is that the growth of the Internet provides us a platform for popularizing telemedicine services with more sophisticated applications possible. There is a need to ensure that what is sent across the Internet is carefully chosen.

Before leaving this introductory chapter, we should also succinctly return to data security. As the Internet is a shared medium, we should be reminded of the risk of security breach as anyone can access the Internet. Telemedicine demands the highest standard of data security, both in terms of information accuracy and patient privacy.

References

Brown, N. (1996), Telemedicine Coming of Age, *Telemedicine Information Exchange*, 28 September, http://tie. telemed.org/articles/article.asp?path=telemed101&article=tmcoming_nb_tie96.xml

Kantor, M. and Irving, L. (1997), *Telemedicine Report to Congress*, US Department of Commerce in conjunction with the Department of Health and Human Services, 31 January, http://www.ntia.doc.gov/reports/telemed/index.htm

Kleinrock, L. (1961), Information Flow in Large Communication Nets, *Research Laboratory of Electronics Quarterly Progress Report*, MIT.

Ledley, R. S. (1965), *Use of Computers in Biology and Medicine*, McGraw-Hill New York.

Licklider, J. C. R. and Clark, W. (1962), On-Line Man-Computer Communication. *AFIPS Conference Proceedings* 21:113–128.

Moore, G. T., Willemain, T. R., Bonanno, R., Clark, W. D., Martin, A. R., and Mogielnicki, R. P. (1975), Comparison of television and telephone for remote medical consultation. *The New England Journal of Medicine*, 292(14): 729–732.

Mullins, M (2001), Exploring the anatomy of a data packet, *TechRepublic on ZDNET*, http://articles.techrepublic. com.com/5100-10878_11-1041907.html

Reid, J. (1996), *A Telemedicine Primer: Understanding the Issues*, Innovative Medical Communications. ISBN: 0965304507.

Stanford, V. (2003), Using pervasive computing to deliver elder care, *IEEE Pervasive Computing*, 1(1), pp. 10–13.

Wang, C. K., Wang, Z., Chen, P., Xie, P., and Hsieh, P. P. (1999), *History and Development of Traditional Chinese Medicine*, IOS Press, ISBN 7030065670.

Zuse, K. (1936), *Konrad Zuse's First Computer: The Z1*. Germany.

2

Communication Networks and Services

Communication networks provide support for a very wide range of healthcare and medical services. Telemedicine uses various types of networks so that physicians can share ideas, surgeons anywhere in the world can perform a single operation together irrespective of where the operating theatre is, nurses and paramedics can retrieve a patient's record anytime anywhere. Hospitals and clinics use the network for everything from patient care to administrative work and inventory management. In this chapter, we learn about the fundamentals of telecommunication technology, with an emphasis on wireless networking, since most telemedicine applications require the flexibilities that wireless networking provides.

2.1   Wireless Communications Basics

To understand how telemedicine works, we must learn about fundamental telecommunications theory. Telecommunications is about delivery or exchange of information between different entities. The most primitive communication example is perhaps two persons talking to each other, where the voice that conveys information is transmitted through the air and reaches the ears of the person who listens. Any communication system would consist of a transmitter (sender), receiver (recipient), and a channel (the path where the information passes through), as illustrated in Figure 2.1. Here is how it works. The transmitter sends out information $s(t)$. The notation $s(t)$ is a function of time meaning the information content varies with time. For simplicity, we may interpret this as the 'sent' information at a given 'time'. This passes through the communication channel and the receiver is presented via the channel with $r(t)$, the 'received' information at a given time. This sounds simple enough. It is logical to think $s(t)$ and $r(t)$ are identical. However, in practice this is almost always not the case.

Unfortunately, the channel causes degradations such as additive noise, distortion, attenuation, etc. Before we proceed further, let's briefly explain what these terms mean. Additive noise is something that is induced to the information and eventually becomes part of the information. In a way, additive noise is added to the original information sent as contamination.

Telemedicine Technologies: Information Technologies in Medicine and Telehealth   Bernard Fong, A.C.M. Fong, and C.K. Li
© 2011 John Wiley & Sons, Ltd

Figure 2.1   Block diagram of a basic communication system

When two people talk, the person who listens may hear other background noise from different sources. Distortion is the warping of the information, causing the information to be altered. It should be noted that the effects of distortion are often considered as some form of noise. Attenuation is the weakening of signal over the distance travelled; the intensity decreases as it propagates away from the sender and may eventually fade out completely. We shall discuss more degradation factors in this chapter. Having established the fact that information received is highly unlikely to be identical to what is sent, let us redraw the basic communication system to that of Figure 2.2, this block diagram shows that noise is added along the channel. This does not necessarily mean noise cannot be induced at the transmitter or receiver. Here, we can write a simple expression to describe the process of communication:

$$r(t) = s(t) + n(t) \tag{2.1}$$

$n(t)$ can take many different forms, the one thing in common is that it will degrade the received information quality. In severe cases, corruption will be so great that the information cannot be correctly interpreted by the receiver. For the sake of completeness a filter is added to remove the noise but its effectiveness can vary significantly in different systems under different situations.

The distance over which information is transferred in a telemedicine system can be as short as a few micrometres within a device or even within an integrated circuit (IC) chip, or thousands of kilometres across continents. The channel can take the shape of copper conductors

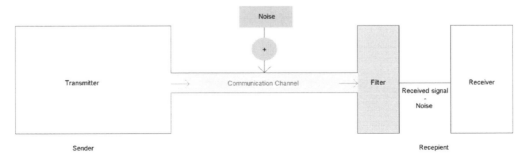

Figure 2.2   Communication system under the presence of noise

having physical connection between the transmitter and the receiver, or 'wireless' over the air. Regardless of what the channel is, maximizing the transmission speed is always of great concern since more information can be conveyed within any given time period. This is similar to operating a bus where the bus company would like to maximize its utilization by having as many passengers as possible; there would be little difference to the operation between carrying 5 and 50 passengers. By the same token, a given communication channel should carry as much information as possible. (Shannon, 1948) describes how noise can affect the maximum transmission speed of a communication channel. We do not intend to go deep into Shannon's information theory, but an extract of his landmark work is worth mentioning to understand the effects on telemedicine performance.

Before leaving this overview of communication systems, it is timely to introduce the term 'transceiver' as it will appear throughout the book. It describes a device that can simultaneously act as a TRANSmitter and a reCEIVER, hence combined together to form the word transceiver. Transmitter and receiver are frequently abbreviated as Tx and Rx, respectively.

2.1.1   Wired vs. Wireless

Wireless communication systems have been gaining popularity as a direct result of technological advancements that have effectively solved numerous reliability and security issues that have traditionally confined the use of wireless technology in low-cost critical applications. Mobility and convenience are undoubtedly driving factors for opting to go wireless. Although both wired and wireless communications are widely used throughout the world, a comparison is given here.

Wired communications have been used for over a century. Following the invention of telegraphy in the mid-nineteenth century, the invention of telephony (Bellis, 2008) began when A. G. Bell and E. Gray worked on the first telephone that made use of a microphone to pick up a person's voice and a speaker that reproduces the voice. The audio signal picked up was transferred through wire that connects two telephones together. This formed the basis of using electric wire for telecommunications. Even before this, wired communication appeared as early as 1794 when C. Chappe started sending telegraphs visually through a line of sight (LOS) communication channel. The meaning of 'line-of-sight' should be quite self-explanatory. It simply means the receiver can 'see' the transmitter, unobstructed. This means that if you sit on the receiving antenna, the transmitter's antenna should be in sight either with the naked eye or through binoculars depending on the distance separating the transmitter and the receiver. However, radio LOS is slightly broader than visual LOS because the radio horizon extends beyond the optical horizon as radio waves follow slightly curved paths in the atmosphere.

Combining the two brought the beginning of optical communications when J. Tindall discovered around the 1870s that light followed a curved water jet as it was poured from a small hole in a tank that subsequently led to the idea of keeping travelling light within a curved glass strand (Hecht, 1999). The works of these inventors formed the basis of wired communication technology that evolved over a century to support a wide range of services. Currently, wired technology is so reliable that it can easily provide at least 99.999% reliability, that is, a failure of no more than 0.001% of the time or less than 5.5 minutes per year. We shall compare the two major types of wires for communication, namely electrical conductors and fibre optic cables in the sub-section 2.1.2.

The commencement of wireless technology dates back to 1887, almost as early as the first telephone, c.f. (Garratt, 1994). This was when D. E. Hughes and H. Hertz began generation of radio waves with a spark gap transmitter. Such underlying technology formed the basis of radio broadcasting by pioneers M. Faraday and G. Marconi at the end of the nineteenth century. Over three decades of the first radio came television broadcasting in the 1930s followed shortly by commercially licensed television stations introduced in Pennsylvania and New York in 1941, long after the first electromechanical television appeared in Germany in 1984 (Sogo, 1994). So far, radio and television broadcast are both one way communication systems, known as 'simplex' communications.

Two-way radio communication was used during World War I but commercial use became popular only after World War II. Although N. B. Stubblefield held a US patent for his wireless telephone in 1908, cellular phones only became widely available from the early 1980s when the FCC (Federal Communications Commission) approved the AMPS (Advanced Mobile Phone Service) system. Up till now the perceived advancement of wireless communications may not be obvious to end users since all these merely let users talk to each other verbally without any added features. '2 G' GSM (Global System for Mobile communications) was launched in Europe in 1991 and has supported text messaging since 1993. Shortly after, 2.5 G and 3 G came with an array of new features such as MMS (Multimedia Messaging Service), video call, Internet surfing, just to name a few. We see how fast wireless technologies have advanced over the past decade or so. It is all about 'speed'. Sub-section 2.1.3 will discuss all we need to know about transmission speed.

So, wired and wireless technologies have evolved over a century and are both very matured technologies by now. A summary of their basic properties are listed in the Appendix. An interesting point to note is that wired and wireless are classified as 'guided' and 'unguided' media, respectively. Figure 2.3 explains the two types. Information travelling across a cable is 'guided' through a fixed path (namely the cable itself), whereas wireless communication

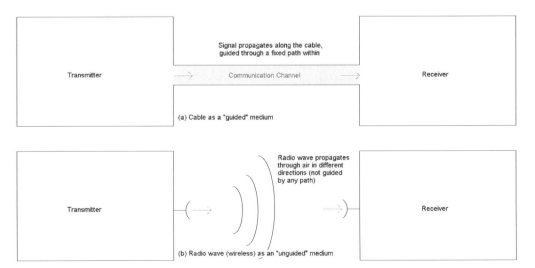

Figure 2.3 Guided versus unguided transmission medium

does not have a fixed guidance for where the information can travel through hence given the description 'unguided'. Briefly summarizing their respective merits and drawbacks in telemedicine, wired communication is more reliable and cheaper for short length deployment, whereas wireless communication provides the convenience of high mobility and deployment flexibility. Wireless communication is a preferred option in most telemedicine applications because of the requirements for mobility: no one wants a clump of wires tangled all over the body!

2.1.2   Conducting vs. Optical Cables

While we have said mobility is an important decisive factor for the dominance of wireless communication in telemedicine applications, it is still important to learn the basic properties of metal conducting cables and fibre optic cables because they are still needed in certain areas, such as network backbone or connection between fixed devices. In this sub-section, we shall look at how these cables convey information, and compare their properties that make them suitable for certain applications.

We briefly discuss the operation of metal conducting cable by looking at a 'twisted pair' cable, illustrated in Figure 2.4. It shows two insulated wires twisted with each other in a helical structure. This is a type of copper cable commonly used in computer and telephone networks. The way they carry information is very simple, a certain voltage represents logic '1' and another voltage level represents logic '0'. The exact representation depends on the specific encoding mechanism used but for the sake of discussion, we may assume a positive voltage denotes a '1' while the lack of voltage (0 V) represents a '0'. In this context, carrying information is simple, the cable simply carries a voltage that alternates between a positive voltage and a 0 V when transmitting a sequence of '1's and '0's.

Optical communications work in a very similar way. Looking at the illustration shown in Figure 2.5, a light beam travels through the centre core when a '1' is transmitted. In contrast, the lack of light represents a '0'. So, the light beam that comes out of the end of a fibre optic cable will be successions of on and off. Of course, the switching is far too rapid to be seen by a human eye hence it may appear as always on. The cable can be bent so there must be some kind of mechanism for retaining the light within the cable's core. Figure 2.5 shows a cladding that surrounds the centre core. It is a highly reflective material that reflects the light back into the core and prevents it from escaping.

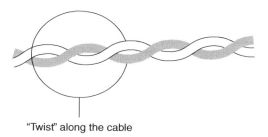

"Twist" along the cable

Figure 2.4   Twisted pair cable

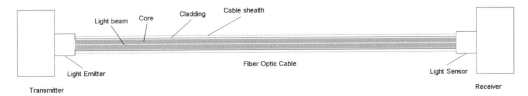

Figure 2.5   Fibre optic communication system

In both cases, '1's and '0's are transmitted across a cable by the presence or absence of a signal. It is important to mention at this point that in practice what goes on behind the scene may not be as simple, but the above discussion does illustrate how transmission is accomplished. Before we leave the discussion on cables, we should look at some major types of commonly used cables in wired telemedicine networks. Another type of metal conducting cable is the 'co-axial cable'. It is no longer popular with telemedicine applications, but warrants a brief mention as this type of cable appears in many places. The most notable situation is TV antennas and decoding boxes. Its main feature is the centre core conductor in much the same structure as fibre optic cable having another group of metal conducting strands surrounding the centre conductor separated by an insulator. The main disadvantage of this type of cable is its bulkiness. There are also other types of wirings like very simply a couple of wires running in parallel. For fibre optic cables, two major types are glass and plastic fibre, with the major difference being a trade-off between performance and cost. In general, the former supports higher transmission rate and is more reliable, whereas the latter is usually cheaper per unit length.

2.1.3   Data Transmission Speed

'Bandwidth', which determines the amount of information a given channel conveys, is a vital term to understand in any aspects of communications. The bandwidth of any given channel is fixed. As a rule of thumb, higher bandwidth supports higher data rate. Since the bandwidth of a given transmission medium is fixed, it may be possible to increase the data transmission rate by stuffing more bits into one 'baud'. Baud is defined as a count of the number of changes of electronic states per second. For example, a copper cable of 1K baud changes the voltage 1 000 times a second. An important point to note is that it does not necessarily mean it only carries 1 000 data bits per second. To illustrate this we will look at some mathematics behind the scene, although we do not intend to go deep into proving the concepts.

In each baud, or a change of signalling state, there is a certain number of different signal levels L, an example would be voltage levels like 0.5 V, 1.0 V. Combinations of binary bits can be assigned to these different levels, e.g. each level represents two bits such that 0.5 V represents '01' and 1.0 V represents '11'. The number of bits n per baud has a simple relationship of:

$$n = \log_2 L \tag{2.2}$$

or:

$$L = 2^n \qquad (2.3)$$

So, in this particular example we have two bits ($n = 2$) and uses four different levels ($L = 4$) each representing: '00', '01', '10', and '11'. By using more different signalling levels, more bits can be carried per baud hence the data transmission rate (or bit rate), measured in number of bits per second or *bps* can be increased for a given fixed baud rate.

Bandwidth is a very important term used when describing the data transmission rate that a given channel supports. It refers to the band of frequencies that an electronic signal occupies when transmitting data across the channel. Therefore, bandwidth of a given channel is measured in hertz (Hz) as often the difference between the maximum frequency and the minimum frequency used. For example, a telephone channel that transmits voice data between 300 Hz (minimum frequency) and 3 400 Hz (maximum frequency) has a bandwidth of 3.1 KHz. So, what is the relationship between the channel bandwidth and data transmission speed?

Nyquist theorem states that the bit rate R_b of a channel of bandwidth H is:

$$R_b = 2.H \ \log_2 \ L \qquad (2.4)$$

This is, of course, the maximum data transmission rate that a channel can theoretically achieve. There are many factors that cause an actual communication channel to have a lower bit rate than this.

Remember, earlier we said that more bits can be carried by each change of signalling state to improve the transmission efficiency by using more different levels. However, having more different levels means signalling levels are squeezed closer together. For example, in the example above each step is 0.5 V. Instead of representing two bits we use eight levels to represent four bits per levels, we may reduce the separation between levels from 0.5 V to only 0.25 V. The single most important problem here is noise that may cause the signalling levels to overlap. The noise level N corresponds to the minimum separation between two levels before the noise causes error to cross the boundary of the adjacent level. The maximum number of levels L is given by:

$$L = \sqrt{\frac{S}{N} + 1} \qquad (2.5)$$

where S is the maximum or peak signal power level. In general, the maximum data transmission rate R_b is directly proportional to peak signal power S, and inversely proportional to channel noise N. A communications system should provide the highest possible transmission rate at the lowest possible power under minimal noise. The above gives us some background theory about data transmission speed.

2.1.4   Electromagnetic Interference

One major drawback of wireless communications is electromagnetic interference (EMI), since EMI effects are far more problematic than with conducting cables. This is particularly risky in healthcare applications because wireless transmitting devices can severely affect the operation of some delicate medical instruments (Tikkanen, 2005) has reported various ways of combating EMI effects in healthcare applications to ensure operational reliability. Amongst various solutions available, providing proper shielding by the use of appropriate housing material for medical instruments can effectively safeguard the device from picking up unwanted interferences. Many composite materials can be used for this purpose. Metallized plastic materials are suitable in housing many types of devices as they can be thermoformed into virtually any shape and are considerably lighter than most metal alloys while providing shielding effectiveness comparable to that of metals. There are three potential problems that cause EMI: source radiating noise, receiver picking up noise, and coupling channel between source and receiver.

All wireless transmitting devices are vulnerable to EMI from nearby radiating sources. These include laptop computers and cellular phones operating in the surrounding area. Such interference usually affects electronic circuitry by causing capacitive coupling, meaning that energy is charged up inside the circuit. Here, a changing electric field is generated that can be capacitively coupled to nearby equipment. There are two major types of EMI, namely continuous and transient interference. The former is caused by emission of radiation consistently by nearby sources such as other transmitting devices or medical instruments. The latter is intermittent such that sources radiate short duration energy. These can be caused by thunderstorms triggering Lightning Electromagnetic Pulse (LEMP) or switching of high current circuits. Standards concerning the regulation of EMI are primarily handled by the International Electrotechnical Commission (IEC), while the Comité International Spécial des Perturbations Radioélectriques (CISPR, directly translates to 'International Special Committee on Radio Interference') deals with radio interference related issues. It is, incidentally, worth noting the 'CE' mark that commonly appears in electronic products including healthcare and medical equipment. 'CE' signifies Conformité Européenne (or 'European Conformity' in English). Products bearing the 'CE' mark indicate conformance of European Directives that require Electromagnetic Compability (EMC) tests to be conducted to ensure a given product complies with the European Union (EU) directive 2004/108/CE before it is permitted to be sold in any member states of the EU.

2.1.5 Modulation

Before concluding our discussion on telecommunication fundamentals we look at the term 'modulation'. It refers to a process where a 'carrier signal', which provides the necessary energy for the information to be delivered to the receiver, is altered in some way according to the information to be carried. This is essentially a procedure of stuffing data into a signal for transmission. The method of altering certain parameter(s) of the carrier signal is changed to represent the data. For example, in FM (frequency modulation) radio broadcast the carrier signal's frequency is changed in relation to the voice information. Such frequency variation would be interpreted by the receiver (radio) as the voice carried over. In its primitive form, parameters that can be changed include the amplitude (the signal level), frequency (number of oscillations per second), or phase (the signal's relative position to time). In more complex modulation schemes, more than one parameter can be changed simultaneously so that more information can be represented per baud thereby increasing the transmission efficiency. In general, the higher the spectral utilization efficiency (SUE), the more the receiver's electronic

Table 2.1 Properties of some common wireless systems

Network Type	Frequency Range	Speed	Maximum Range
Bluetooth	2.4–2.485 GHz	3 Mbps	300 m
IR	100–200 THz	16 Mbps	5 m
Wi-Fi	2.4–5 GHz	108 Mbps	100 m
ZigBee	900 MHz	256 Kbps	10 m
Cellular Networks	850–1900 MHz	20 Mbps	5 km
WiMAX (Fixed)	10–66 GHz	1 Gbps	10 km
LMDS	10–40 GHz	512 Mbps	5 km

circuit structure complexity is required as resolving between different possible states of the signal becomes more difficult. SUE is a measure of how efficient a modulation scheme is to carry a certain amount of data for a fixed bandwidth.

2.2 Types of Wireless Networks

Wireless communications have been developed to such an extent that numerous options exist. Different network types are optimized for different applications, with coverage ranging from a few metres to thousands of kilometres. In this section, we introduce some commonly used networks in telemedicine applications and explain why they are suitable for specific situations. Key properties are summarized in Table 2.1.

2.2.1 Bluetooth

This technology provides short range coverage primarily for mobile devices connected in an ad hoc network called 'piconet' within a room. Key selling points are low cost requiring simple circuitry and low power consumption. Its flexibility for connecting between devices in close proximity together may pose a threat of spreading computer virus. Bluetooth uses adaptive frequency hopping (AFH) to reduce EMI by detecting other devices in the spectrum and hops between 79 frequencies at 1 MHz intervals, so as to avoid the frequencies nearby devices are using. Bluetooth technology is overseen by the Bluetooth Special Interest Group (SIG). There are currently three classes covering around 3 m, 30 m, or 300 m.

Although it is widely seen in hands-free units of cellular phones, it is useful for small wearable biosensors due to low power (1 mW for 3 m or 10 ft, Class 3) and simple, low cost transceiver.

2.2.2 Infrared (IR)

Infrared waves sit between microwave and visible red light of the spectrum. A considerable amount of infrared radiation is emitted from the sun and is usually associated with heat. In fact, approximately an equal amount of infrared and visible light hits the earth's surface from the sun. So, what does it have to do with communications and healthcare? On a related front, infrared detection is widely used in night vision, which is imperative in search and rescue. A popular example of infrared in wireless communications is remote control for home appliances. When we pick up the remote control to adjust the volume of a stereo, the controller emits an infrared signal that carries the instruction to the stereo's sensor.

Infrared is classified into three different categories by the International Commission on Illumination (CIE), of which near-infrared or IR-A is used in night vision applications; whereas wireless communications usually use short-wavelength infrared (IR-B). It is worth noting that short-wavelength infrared is widely used in long range optical communications although we will not go into the details here as we are looking at wireless networking. IR wireless standards are governed by the Infrared Data Association (IrDA) for devices that use the successive 'on' and 'off' of an infrared light emitting diode (LED) for communication. At the receiver, a silicon photodiode converts the received infrared pulses to an electric current replicating the sequence of 'on' and 'off'. It is a very mature technology used for decades and very easy to implement with virtually no interference issues although it does not have the ability to penetrate through walls. Another major issue is that it requires direct LOS and the transmitter must be aligned fairly close to the centre of the sensor with only +/- 15° offset possible. Although current IrDA compatible devices support only up to 16 Mbps, the introduction of Giga-IR offers a theoretical speed of up to 1 Gbps. It is often used in small ECG fragment transmission.

2.2.3   Wireless Local Area Network (WLAN) and Wi-Fi

The IEEE 802.11 standards are very widely used in wireless home networks, providing a low cost and convenient way for Internet access. Unlike Bluetooth and IR, WLAN requires some efforts in setting up initial configurations before a communication link can be established. Popular IEEE 802.11 standards include a/b/g/n; these standards define the specifications for the physical layer ('PHY' which defines how raw data bits are transmitted over the air) and the WLAN's Medium Access Control Layer ('MAC' which provides addressing and channel access control procedures that allow several devices to communicate with a single Access Point). The standards provide details of these layers so that devices can be designed with full compliance to ensure operability. Apart from 802.11a which operates at 5 GHz, the remaining three standards are 2.4 GHz. In this frequency band, significant interference can be caused by appliances of similar frequencies such as cordless phones, microwave ovens, and also Bluetooth devices. Its coverage varies greatly depending on whether it is used for indoor or outdoor operation, ranging from 50 to 300 m, respectively.

A basic WLAN consists of at least one access point (AP) and the mobile client(s) (MC), the MCs are essentially any mobile devices that seek to maintain a wireless connection to the network via the AP. APs are placed in various locations throughout the coverage area to form the wireless network infrastructure. In its most primitive configuration, there is one AP in the centre and one or more MCs operating around it. The network coverage area can be increased by installing more APs. A wireless relay can be installed to further extend the range. When there are multiple APs within the proximity an MC selects the closest AP whose signal strength is strongest for communication.

Information security is always a great issue due to its popularity and sharing of the unlicensed ISM band. The details will be covered in Chapter 6.

Wi-Fi provides a unified standard derived from IEEE 802.11 WLAN by the Wireless Ethernet Compatibility Alliance (WECA) for different types of wireless devices. Wi-Fi is sometimes referred to as Wireless Internet. Wireless devices are served by an access point, also known as 'hotspot'.

Both Wi-Fi and Bluetooth have many similarities, but there are also many differences due to tradeoffs between coverage, data speed, and power consumption hence device size and cost. Due to its popularity in home networking, Wi-Fi technology is very commonly used in off-site patient monitoring for people recovering at home as existing home networks can be utilized with minimal alternation.

2.2.4   ZigBee

These are small digital devices for wireless personal area networks (WPANs) complying with the IEEE 802.15.4 standard. Easy to implement and very low power consumption, they are not intended for intensive data transfer due to their slow speed and are primarily used for wireless control and monitoring. Currently there is no global standard operating frequency: 868 MHz in Europe, 915 MHz in USA, 950 MHz in Japan, and 2.4 GHz in most other parts of the world. In a way it may be viewed as a simplified version of Bluetooth and is often use in System On Chip (SoC) implementations. It is so cheap that a transceiver is available for below US$1 per unit and is regularly used in safety precautionary devices such as smoke detectors and air conditioning control. It is also used in body area sensor networks (see Section 3.5 for details). Communication network is served by a Zigbee Co-ordinator (ZC) and access through a Zigbee Router (ZR) which effectively relays data between devices.

2.2.5   Cellular Networks

Mobile phone networks are commonly known as cellular networks because the coverage area is composed of radio cells each served by a base transceiver station (BTS). Functions of the BTS may diverge considerably as determined by the service operator and the cellular technology. Coverage area can be enhanced by the establishment of more cells. In addition to improving coverage, the use of cellular composition also expands capacity and lowers transmission power requirements. The ability of users moving across cells with continuous connection is one of cellular network's key features, provided by 'handover' algorithms. There are different technologies currently used in different parts of the world. We will give a brief account on those still widely used today, while omitting obsolete systems such as Advanced Mobile Phone System (AMPS) and Time Division Multiple Access (TDMA) cellular technologies.

CDMA1900 (1.9 GHz): Stands for Code Division Multiple Access 1.9 GHz. There is an old digital cellular communication system still used in the USA, as some operators are licensed to operate at 800 MHz as legacy systems rolled out prior to the FCC approval for 1.9 GHz. CDMA supports multiple simultaneous base stations on the same frequency channel.

2.5 G (900 MHz): GSM Phase 2+ (Global System for Mobile communication) as defined by the European Telecommunications Standards Institute (ETSI). A system widely used in most parts of the world offering ease of roaming across countries with one single cellular phone. GPRS (General Packet Radio Service) is an extension of 2.5 G that supports a wide range of multimedia services at fairly slow speeds of up to 114 Kbps. The type of services supported is governed by an Access Point Name (APN) defining services such as Wireless

Application Protocol (WAP) access, Short Message Service (SMS), Multimedia Messaging Service (MMS), Point-to-point (PTP) as well as Internet access.

3 G (1.8 GHz): Third Generation technology improving upon the previous 2.5 G with a maximum speed of 14.4 Mbps. Main features of 3 G include video calling and mobile TV broadcast. There are different interface systems defined by the ITU (International Telecommunications Union) IMT-2000 as 3 G networks. The most significant ones are Mobile WiMAX and UMTS (Universal Mobile Telecommunications System), which is also known as W-CDMA where W denotes Wideband. The former is named under 'Worldwide Interoperability for Microwave Access' and is developed from the IEEE 802.16 Broadband Wireless Access (BWA) standard (see below) whereas the latter is a much more mature and widely used technology whereas the latter is a direct successor to 2.5 G that basically evolves from previously available mobile technology. An improved version, widely referred to as 3.5 G launched in 2006, is High Speed Downlink Packet Access (HSDPA) that supports over 20 Mbps. This is expected to evolve to 4 G at over 100 Mbps with enhanced security around the year 2012.

Between 2.5 G and 3 G, there are popular technologies often classified as '2.75 G', a term not often seen, but the following expressions should be very familiar to readers: CDMA2000 and EDGE (Enhanced Data rates for GSM Evolution). These are developed from CDMA1900 and GSM Phase 2+. There are misconceptions of these being classified as 3 G because of their increased data rates supported versus 2.5 G systems.

PHS (1.9 GHz): Personal Handyphone System used exclusively in Japan for its high portability resulting from low power consumption and lack of a SIM card. The system was mainly designed for voice calls with data support of up to 256 Kbps. PHS is progressively replaced by 3 G networks.

2.2.6   Broadband Wireless Access (BWA)

BWA supports a diverse variety of services due to its ultra high speed. Usually used for medium to long range distribution, the carrier frequency can be anywhere from a few GHz to 40 GHz depending on local regulations. Development of BWA is governed by the IEEE 802.16 Working Group on Broadband Wireless Access Standards. Note that IEEE 802.16 does not specify frequency bands or equipment certification requirements. WiMAX, fixed or mobile operating in 2.4–5 GHz ISM band, is compliant with both IEEE 802.16e and ETSI HiperMAN wireless Metropolitan Area Network (MAN) standards covering tens of kilometres, is becoming very popular over recent years with a high degree of interoperability primarily. Local Multipoint Distribution Service (LMDS), being a prevalent BWA deployment, is intended for fixed network deployment implying that mobility support is very limited. The main difference between Fixed WiMAX and LMDS is the operating frequency that leads to a significant increase in channel bandwidth. LMDS is capable of supporting over 512 Mbps for carrying vast amounts of data. Since the radios have 90° field of view, it is possible to set up four radios for an omni-directional 360° coverage.

The properties of LMDS make it particularly suited for telemedicine backbone support. The term 'backbone' refers to the medium that provides a main trunk line for interconnecting different local area networks (LANs) as well as equipment together over a large area. For example, a hospital may have several buildings having a network backbone in a hospital interconnecting different entities together as shown in Figure 2.6.

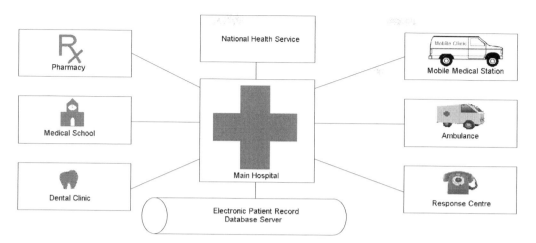

Figure 2.6   Network infrastructure linking the hospital to many supporting entities

2.2.7   Satellite Networks

These are sophisticated and expensive networks as placing a satellite precisely above the earth is a very costly exercise. Its operating principle is, however, reasonably straightforward. A communication satellite ('comsat') is laid in a pre-determined orbit above the earth. The choice of orbit depends on the desired coverage area. The comsat serves as a point-to-point microwave radio relay that provides a radio link between two earth stations. Satellites are frequently used in wide area networks (WAN). Despite being vulnerable to environmental interference such as solar storms, it is very reliable and provides very high speed links. Although such properties may appear suitable for remote robotic surgery given the vast amount of data that needs to be transmitted, its inherent long propagation delay will likely affect real-time operations. For this reason, satellite networks are mainly used for remote recovery.

2.2.8   Licensed and Unlicensed Frequency Bands

From the above discussions we learned that some networks operate in licensed bands while others are unlicensed and shared with many other users. So, what are the differences that might affect telemedicine operations? Both licensed and unlicensed-band equipment can operate co-operatively for any telemedicine application, finding out which is a better choice may depend on different situations (Dekleva, 2007). First, an unlicensed network does not incur any implementation delay and cost on acquiring a license, an unlicensed connection can be easily established and anyone can do it with no restriction on the type of radio device used. Access by anyone means devices are at risk of security breaches and interference. Licensed networks operate within designated bands with exclusive use so equipment can be highly customized to exact requirements. Interference protection (there is no assurance of an interference free environment) and guaranteed bandwidth availability are key features of using licensed frequency bands. So, generally there is a compromise between cost and convenience versus security and operating environment.

By now we have looked at several different types of wireless networks that can be utilized for a wide range of telemedicine applications. Each has its own advantages and disadvantages. Careful selection based on their performance and properties will ensure that the type of services possible can be vast. Very often, the use of an existing network is a desirable choice due to cost effectiveness and reduction in implementation time. As communication technologies advance and more choices become available in the near future, telemedicine is set to become even more reliable and accessible to more people of different needs.

2.3   The Outdoor Operating Environment

As signal strength weakens over distance travelled (attenuation), the effects of electrical noise radiated by other nearby devices can be very significant. The noise can be so severe that the transmitted signal can be lost or corrupted, causing the received data to be useless. In addition to noise and attenuation, signal distortion can be an issue as it travels through metal conductors. Distortion can take various forms depending on what lies along the signal path, but generally speaking the signal's shape is distorted, for example, when a square wave no longer retains its smooth pulse. Although these signal propagation issues also exist indoor, there are more uncontrollable factors in the outdoor environment that make many signal degradation factors more severe.

The benchmark by which signal loss is measured in a transmission link is the loss that would be expected in free space, i.e. the loss that would occur along a path that is free of anything that might absorb or reflect signal energy. When a propagating radio wave hits a physical obstacle, it is subject to the following phenomena, illustrated in Figure 2.7:

Diffraction: the signal splits into secondary waves. It happens when the propagating signal hits a surface that has sharp edges. The waves produced by the surface are present throughout space and a certain portion may penetrate behind the

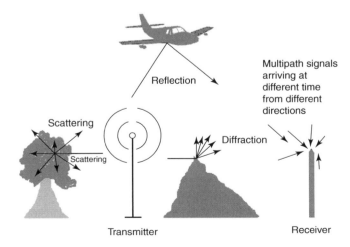

Figure 2.7   Propagating wireless signal degrades due to different phenomena

obstacle thereby causing power loss, giving rise to the bending of waves around the obstacle.

Reflection: the signal reflects back to the transmitting antenna, just like light being reflected by a mirror. It happens when the propagating wave hits a physical object that is much larger than the wavelength of the carrier.

Scattering: the signal reflects with different components spread in different directions as being diffused upon hitting an obstacle. Contrary to diffraction, scattering is an issue when the object that the propagating wave hits is small compared to the wavelength, such as rough surfaces, dust and air pollutant particles, or other irregularities in the channel. When the signal scatters in all directions, it effectively provides additional energy as perceived by the receiver. This leads to the actual received signal being stronger than that affected by reflection and diffraction.

All these will result in loss of signal strength, collectively known as *fading*. Such an effect can be compensated for by using multiple antennas to pick up different components of the same signal arriving from different directions, such technique is known as 'space diversity'. How it works is very simple, since different components are subject to different time delay, phase shift, and attenuation. One antenna may experience too severe fading that cannot effectively pick up the signal. Whereas the use of more antennas will improve the chances of picking up a better version of the same signal.

In outdoor signal propagation, very often the signal must clear large physical obstacles such as buildings and trees. One thing to remember is that a visual line-of-sight observed, when looking from the position of an antenna towards another antenna, does not necessarily imply a radio line-of-sight also exists especially in long range communication. Whether this is true depends on the clearance of the *Fresnel zone*, this is because a radio wave needs some space to reach the receiving end as it is obvious that a wave cannot 'squeeze' through a small hole drill in a wall. *Fresnel* zone is defined as a long ellipsoid that stretches between the two antennas. The first Fresnel zone is defined as the spheroid space encircled within the trajectory of the path when the difference between the straight line directly drawn between the two antennas, and an indirect path that crosses a single point at the edge of the Fresnel zone with half the wavelength. This is a spheroid space necessary for the wave to propagate towards the receiving antenna centred along the direct straight line path between the antennas.

For example, suppose the signal frequency is 30 GHz. By applying the familiar formula (Equation 2.6) learned in high school physics:

$$v = f.\lambda; \qquad \lambda = v/f \tag{2.6}$$

Given that the speed of radio wave propagating through free space is approximately 3×10^8 m/s, the wavelength λ would be $3 \times 10^8 / 30 \times 10^9 = 0.01$ m or 1cm. So, half wavelength will be 5 mm. So, the wave reaches the receiver by the direct straight line path, and it also reaches there within a spheroid area of 5 mm. It is said that at least 60% of the first Fresnel zone should clear any physical obstacle in order to achieve propagation characteristics comparable to that of free space. Also, the terrain profile around the spheroid area needs to be taken into account to estimate the path loss or attenuation. This can often be estimated using well-established models such as the Longley-Rice Model (Hufford, 1999), where the median transmission loss is predicted using the path geometry of the terrain profile and the refractivity of the troposphere. An Urban Factor (UF) then accounts for additional attenuation due to urban clutter surrounding the receiving antenna. The model is effective as an irregular terrain model (ITS). However, it does not take into consideration the effects of buildings and foliage. When optimizing the propagating path for long range communication, usually when exceeding five to eight kilometres), the Earth's curvature also needs to be taken into consideration.

The transmission loss depends on how much power reaches the receiving antenna. Attenuation is always an important consideration as the signal will eventually become too weak to be picked up by the receiver. Weather conditions such as rain, fog or snow can severely affect the range and reliability of wireless systems. The effect of rain induced attenuation can be very severe especially in tropical regions where consistent heavy downpour in excess of 100 mm/hr can last for hours. The dB/km measurement of attenuation indicates the power loss in dB for every kilometre of distance travelled. The actual impact is determined by several factors, primarily the rain rate and carrier frequency. In general, the heavier the rain and/or the higher the frequency, the more power is lost per kilometre. As a general guideline, rain induced attenuation is not a significant problem for systems operating under 10 GHz, or when the rainfall rate is below 20 mm/hr. To see how severe this problem is, we take a look at Figure 2.8, which compares the attenuation for 10 and 50 GHz. Note, incidentally, that the plots evaluate vertically polarized signals; signals of horizontal polarization always undergo a higher degree of attenuation than vertical polarization under identical conditions. The difference between two polarizations also increases as the rainfall rate and/or frequency increases as shown in Figure 2.9. The effects of heavy rainfall on radio propagation path reduce the system availability because rain causes cross-polarization interference that subsequently reduces the polarization separation between signals of vertical and horizontal polarizations as the signals propagate through rain. The extent of radio link performance degradation is measured by cross polarization diversity (XPD) that is determined by the degree of coupling between signals of orthogonal polarization, (Fong, 2003a). (Bahlmann, 2008) gives a comprehensive definition of XPD as a measure of the strength of a co-polar transmitted signal that is received cross-polar by an antenna as a ratio to the strength of the co-polar signal that is received, which typically results in a 10% reduction in coverage due to cell-to-cell interference. While it may make

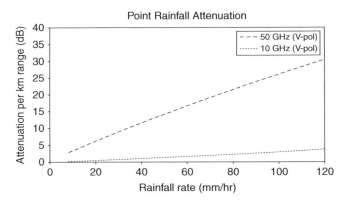

Figure 2.8 Effect of rain attenuation at different rainfall rate

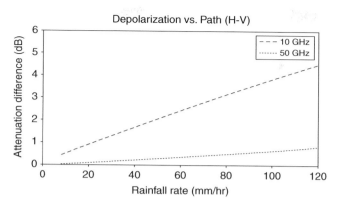

Figure 2.9   Horizontally polarized signal undergoes more severe attenuation than signal of vertical polarization under identical conditions

logical sense to relinquish using horizontally polarized signals to avoid excessive power loss, we shall see in section 3.1 why in many systems both are used simultaneously.

The issue of channel degradation due to rain must be thoroughly addressed in telemedicine because very often accidents occur as a direct result of heavy rain. So, telemedicine systems that assist emergency rescue operations must maintain an adequate level of quality. Optimizing the appropriate system margins would maximize the availability of radio link in such situations (Fong, 2003b).

Multipath fading, a phenomenon resulting from multiple components of a signal reaching the receiver from different directions of arrival (DOA) at different times due to reflection through different physical obstacles along the propagation path subject to different amount of delay is illustrated in Figure 2.10. The shortest path between the transmitter and receiver is the

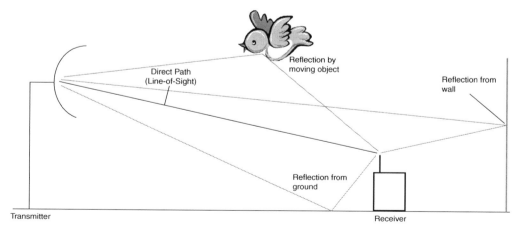

Figure 2.10   Multipath fading caused by different components of the signal arriving at different times through different propagating paths

straight-line unobstructed path having LOS. When the propagating signal hits an obstruction, it will be spread out to take multiple paths resulting in different travelling times to arrive at the receiver causing varying amounts of time delay. Multipath is generally an issue with signals below 10 GHz whereas attenuation caused by rain is the most important consideration factor at frequencies above 10 GHz. Lower frequencies are therefore preferred in tropical regions where heavy and persistent rainfalls are expected. Otherwise systems at higher frequencies operate in a less congested part of the spectrum with more available bandwidth.

Another potential issue with wireless communications that may cause delays is Doppler Spread, where fluctuations are caused by the movement of the transmitter, receiver, or the many physical objects between them. Doppler Spread is particularly relevant in vehicular communications where fast movement can severely affect signal reception.

2.4 RFID in Telemedicine

RFID (Radio Frequency Identification) is an old technology that appeared as early as in WW II but has only been widely used in many applications for everyday life over the past decade or so. As its name suggests, RFID is all about identifying an object using radio frequency signals. So, it is often perceived as an 'electronic barcode' system. There are many different forms of RFID currently used throughout the world. Essentially RFID involves 'tags' that identify an object and 'readers' that read and identify the tags. These come in various forms, portable or fixed readers, active or passive tags meaning whether an internal battery is needed to power a given tag in order to respond to a reader. The battery serves the sole purpose of providing a longer read range that allows an active tag to be read from a longer distance. So, how does a passive tag respond without a battery since it does not have any power source? The answer is actually quite simple as it gets the necessary power from the reader while it receives the reading signal from the reader. Such a signal, carrying a certain amount of energy with it, hits the coiled antenna inside the tag thereby induces a magnetic field that energizes the electronic circuit containing information embedded within the tag including a unique identification number.

Advantages of passive tags are obvious; they are extremely small, cheap, and durable. These tags can be manufactured for less than 10 US cents each and can be mass produced because a tag merely consists of a printed antenna and a small chip wrapped in paper. Figure 2.11

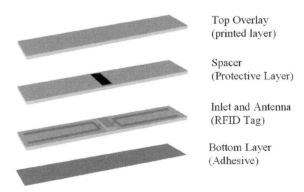

Top Overlay
(printed layer)

Spacer
(Protective Layer)

Inlet and Antenna
(RFID Tag)

Bottom Layer
(Adhesive)

Figure 2.11   An RFID tag

sketches the typical layout structure of an RFID tag. However, short reading range may not be the only problem in telemedicine applications. It does not have the ability to supply power to biosensors if one was to be attached to the tag.

Major problems with RFID reliability are tag collision and reader collision. Tag collision occurs when multiple tags are energized by a single reader so that the tags respond at the same time causing read failure, whereas reader collision refers to situations when within a certain coverage area of one RFID reader overlaps with that of a nearby reader. Another major issue is lack of security since tag signals can be picked up by any reader within range.

RFID systems operate in a number of different frequency ranges. Their propagation properties can severely affect the operation in different telemedicine applications. With LF (low frequency: 135 KHz) and HF (high frequency: 13 MHz) systems, signal reflection severely reduces the transmitted signal power as shown in Figure 2.10. UHF (ultra high frequency: 900 MHz) systems can suffer from signal absorption by water making them unsuitable for applications involving placements of tags on a human body.

RFID finds its use in many medical applications. To name a few, it is very widely used in drug dispensary, linking patients with restricted or controlled drugs. Tracking babies and other patients as well as medical equipment is also one key area of RFID usage. The list of applications is seemingly endless. An area that warrants more thorough discussion is implantation with medical devices within a human body. This is a challenging yet important application for devices such as Biventricular Pacemaker and Glucometer. UHF is not suitable due to composition of water in human tissue. Due to lack of security features and high cost readers, HF is a clear choice for surgical implantation. To implant a biventricular pacemaker, leads are implanted through a vein into the ventricle and the coronary sinus vein to regulate the ventricle. Since it is intended to serve patients suffering from serious heart failure symptoms, any irregularity must be reliably reported via telemedicine network without delay in order to minimize the risk for sudden cardiac consequences. Further, patients with inadequate ejection fractions may require an implantable cardioverter defibrillator (ICD) in conjunction with the pacemaker to ensure sufficient heart pumps per beat are maintained. Since an ICD functions by rhythm detection and shocking the heart, such action can affect the operation of an associated RFID tag. Also, each tag associated with an implanted device may risk tag collision as they are so closely placed to each other. Obstacles along the signal propagating path include the lung's anterolateral surface of the inferior lingular segment, followed by rib bone and finally all the way through skin that consists of epidermis, dermis, and subcutaneous fat leaving the body at the chest. There are many layers of barriers that would affect the signal path.

Capacitance between a tag and its housing can severely impact the antenna's tuning. To combat this problem, a tag must be tuned away from resonating at the reader's frequency to diminish mutual coupling with the other tag. The read range can therefore be improved by using RFID tag with tunable antenna.

The above case study may sound rather complicated. Let's look at a less demanding example of implantable glucose meter for diabetes monitoring documented by (Carlson, 2007), which does not have any immediate life threatening consequences in the event of communication failure. The RFID tag would be responsible for transmitting the glucometer reading away from the body for subsequent analysis. Since data storage capacity of the tag is no more than 2 KB as with any typical passive tag, the data needs to be sent away as soon as it is received from the glucometer before it becomes full. Here, the part of the RFID tag is rather similar to a cellular phone connected to a laptop computer as a wireless modem via a USB (Universal Serial Bus) cable. In this analogy, the mobile phone acts as a point of sending data to the outside world.

Having understood the tag's function, we need to look deeper into the mechanism involved. As a wireless transmitting device, it acquires the necessary energy from the incoming wave radiated from the external reader; the received energy must be sufficiently strong to power up the chip. When the signal is sent back from the tag's antenna, it must be efficient enough to ensure that the transmission power is adequate for the reader. So, here is the challenge: on one hand we need to ensure the tag is implanted as closely to the person's skin as possible to minimize the signal propagating distance. On the other hand, we also need to avoid immediate contact between the antenna and any internal tissue that may severely shield off the signal. Any housing for the tag that seals it off to avoid direct contact with tissue will certainly have an effect on signal penetration. The choice of material therefore becomes a critical issue in this scenario. (Friedman, 2001) describes various materials suitable for implantation. Polyvinylchloride (PVC) insulator of approximately 10 μm is suggested to be an optimal wrap for providing a reasonable separation between the tag and surrounding tissue without significant impact to signal propagation.

So the communication aspect is more or less resolved, but what about integration with the glucometer? The system does not seem to have many components but the technical issues can be quite challenging as it entails biocompatible interface, glucose sensing and a device to convert the captured reading into an electrical signal that can be written into the RFID tag for subsequent transmission away from the human body to the reader. We also mentioned that mechanisms must be in place to ensure that the previously stored data must be sent away and the tag's memory content emptied before the next set of reading comes in. So, the device that connects the glucose sensor to the RFID tag must be capable of programming the tag's memory in addition to generating the signal from the captured reading. Also, this device must be very small and power consumption must be so low that one energization activity by the reader can produce and store sufficient energy to last until the next energization, i.e. the next read operation. Ultimately, what needs to be achieved is to download the collected data for subsequent analysis and storage. This is mainly determined by the optimal design of an efficient antenna and related circuitry for the chip so that the data can get through to the outside world from inside the body.

RFID is capable of more than acting as an identifier, it is also a very small and economical implantable object that supports short range wireless communications. It is so versatile that its application area is virtually unlimited, it is certainly an important tool for telemedicine.

References

Bahlmann, B. and Ramkumar, P. (2008), XPD - Cross Polarization Discrimination, http://www.birds-eye.net/definition/acronym/?id=1151878664

Bellis, M. (2008), about.com: http://inventors.about.com/od/bstartinventors/a/telephone.htm

Carlson, R. E. and Silverman, S. R. (2007), Development of an Implantable Glucose Sensor, http://www.verichipcorp.com/files/GLUwhiteFINAL.pdf

Dekleva, S., Shim, J. P., Varshney, U., and Knoerzer, G. (2007), Evolution and emerging issues in mobile wireless networks, *Communications of the ACM*, 50(6):38–43.

Fong, B., Rapajic, P. B., Fong, A. C. M. and Hong, G. Y. (2003a), Factors causing uncertainties in outdoor wearable wireless communications, *IEEE Pervasive Computing*, 2003, 2(2):16–19.

Fong, B., Rapajic, P. B., Fong, A. C. M. and Hong, G. Y. (2003b), Polarization of received signals for wideband wireless communications in a heavy rainfall region, *IEEE Communications Letters*, 7(1):14–15.

Friedman, C. D. (2001), Future directions in biomaterial implants and tissue engineering, *Archives of Facial Plastic Surgery*, 3(2):136–137.

Garratt, G. R. M. (1994), The early history of radio: from Faraday to Marconi, *Institution of Electrical Engineers and Science Museum*, History of Technology Series, London.

Hecht, J. (1999), *City of Light: The Story of Fiber Optics*, Oxford University Press, New York.

Hufford, G. (1999), The ITS Irregular Terrain Model, National Telecommunications and Information Administration, Boulder: http://flattop.its.bldrdoc.gov/itm.html

Shannon, C. E. (1948), A Mathematical Theory of Communication, *Bell System Technical Journal*, 27:379–423.

Sogo, O. (1994), *History of Electron Tubes*, IOS Press.

Tikkanen, J. (2005), Wireless Electromagnetic Interference (EMI) in Healthcare Facilities, *BlackBerry Research White Paper*.

3

Wireless Technology in Patient Monitoring

In Chapter 2, we learned that many alternative types of wireless networks are currently available for telemedicine services. These networks have very different properties and are designed for different situations. There is no simple answer as to what type of network is best for telemedicine as different applications may have very different needs. Having looked at a variety of technologies, we have seen propagation as being one major issue that all wireless networks face. We discussed why wireless telemedicine is far more popular than wired systems. Wireless networking is the underlying technology that enables the connection of healthcare in terms of both people and resources, technological advancements over decades have enabled secure and reliable networks to provide services for life-critical services. In this chapter, we look at various situations where wireless telemedicine helps patient recovery and rehabilitation. We shall see how these can be accomplished and what technical challenges exist. The popularity of RFID in numerous applications paved the way for people and medical resources to be easily tracked down and monitored.

Specific network design is motivated by the application it provides such that it needs to fulfil the requirements to reliably transmit the type of information involved. For example, to monitor a ventricular tachycardia (VT) patient requires regular transmission of ECG and heart rate information to ensure that any risk of ventricular fibrillation will be promptly detected; this may require resolving QRS complexes separation of at least 0.05s. In any telemedicine system, we must ensure that the communication network used is capable of supporting the required data rate.

In this chapter, we begin by looking at technologies and challenges of setting up a body area network (BAN) that is suitable for implementation on both patients under monitor and health professionals who serve them. We shall then look at some major applications of remote patient monitoring utilizing wireless communication technology. Readers should be reminded that the examples given may have alternative solutions, these are by no means the only deployment option. The primary objective of looking at these examples is to grasp a good understanding about how telemedicine technologies support rescue missions in different situations and what challenges are faced.

Telemedicine Technologies: Information Technologies in Medicine and Telehealth   Bernard Fong, A.C.M. Fong, and C.K. Li
© 2011 John Wiley & Sons, Ltd

3.1 Body Area Networks

Body Area Network (BAN), also known as Personal Area Network (PAN), is made possible in recent years when technology allows very tiny radio transmitting devices to be securely installed on a human body. In addition to its increasingly popular deployment in healthcare, BAN is also used in many computing and consumer electronics applications due to its flexible deployment options. These devices are so small that some can even be implanted inside the body. It provides the underlying technology for monitoring various signs of the body and to automatically issue an alert should an abnormal behaviour is detected. It also provides a convenient means of logging daily activities and determining whether a user has met a pre-determined target for workout during a session. It therefore supports people who require medical attention or simply for fitness monitoring. Biosensors are attached to the user's body for remote health monitoring offering extremely high mobility; a BAN typically consists of two major components:

> *Intra-BAN* for internal communication around the body, where sensors and actu-ators are connected to a mobile base unit (MBU) that serves as a data processing centre. The MBU can be just about any consumer electronics device that we carry on a regular basis, like a cellular phone, in-car hands free kit, or the wireless modem that we use for connecting our laptops to the Internet.

> *Extra-BAN* for external communication between components surrounding the body and the outside world. This is normally a telemedicine system that conveys the collected data for processing and analysis.

In general, BAN devices have properties of very low power consumption, usually below 10 mW and a low data throughput of around 10 Kbps. There are a number of issues that BANs have. First, data security is a particular issue since no data protection mechanism is employed in most situations. QoS (quality of service) assurance for an individual device must be provided to ensure that all devices remain contacted. Coverage does have not to be vast, typically anywhere within two metres from the BMU should suffice. Antenna design as an integral part of a wearable sensor can be a very challenging task since it needs to provide omni-directional coverage to ensure a high degree of mobility and effects of absorption by human body on signal propagation need to be thoroughly investigated (Hirata, 2010). This is a particularly demanding mission for implanted devices.

Although there are currently no standards for BAN implementation, the IEEE 802.15 Working Group for Wireless Personal Area Networks (WPANs) has been working on, allowing a broad range of possible devices to interoperate on various transmission media. (Li, 2008) has described a number of prospects that may eventually lead to the standardization of IEEE 802.15 for BAN deployment. Different groups are in place for different media. For example, most popular ones are IEEE 802.15.1 for Bluetooth and 802.15.4 for Zigbee.

Due to the high degree of flexibility for different sensors to be attached, BAN is capable of monitoring suffers of asthma, diabetes, heart problems, etc., logging and tracking of related data can be easily accomplished to detect any potential issues. In areas such as hospitals and clinics, where many patients may be monitored in close proximity, one major design challenge to overcome is the ability to distinguish each BAN system associated with each patient so that data collected will not be mixed up. Although many BANs can be built upon off-the-shelf sensors, a number of concerns exist for the sensors:

- *Standards*: Functional specifications, operating environments, communication protocols, operating range, security and privacy.
- *EMC*: Amount of electromagnetic radiation induced, susceptibility to interference.
- *Calibration*: Procedure and frequency for calibration, precision.
- *Integration*: Connections, database linkage, mounting and placement.

Let us go further into these design issues. Currently, there are no standards that govern the development of BAN biosensors, guidelines on power source requirements and communication protocols specifying how data is transmitted do not exist. Performance and reliability of sensors differ when used in different situations. For example, implantable sensors may not be suitable for operating above a certain altitude or when the person is submerged in water while participating in activities such as swimming and boat repairing. How far can a person move away from the point where data is collected needs to be specified to ensure that data can be successfully collected if the device does not have any internal data storage buffers. As with almost all healthcare systems, data security and privacy is an important topic to address. This will be covered in details in Chapter 6.

EMI compliance is applicable to all wireless transmitting devices in most countries and different countries may have different regulations. In cases where health monitoring devices can be brought to different countries, they must be designed to comply with all relevant regulatory governance concerning EMC. Calibration is an important process for all precision instruments in ensuring that the data captured is within the specified accuracy limits. It is possible to incorporate self-calibration and diagnostic functions for ease of maintenance. When this cannot be accomplished, there will be a need to specify how frequent calibration is necessary to maintain accuracy; and whether calibration can be performed by the user. Finally, how each sensor is connected to the MBU and how it is securely installed on the user must be thoroughly addressed to ensure reliability. Sensors can be implanted inside a human body given the appropriate protective housing; many of them are attached to the body on a temporary basis, while some are embedded on clothing as investigated by (Park, 2003) and (Winters, 2003). To ensure maximum mobility, the sensors must be lightweight with small form factor (physical size and shape); the weight and form factor is primarily determined by the internal battery installed within. These sensors must therefore be designed to be exceptionally power efficient to minimize size and maximize durability. Also, frequent replacement or charging of battery would make usage inconvenient and impractical.

To better understand how BAN operates, we look at an example in Figure 3.1 which shows the infrastructure of a basic BAN that consists of sensors for monitoring a cardiac patient under supervised recovery. Sensors are present for collecting ECG data, oxygen saturation, motion sensing for gait phase detection, body and ambient temperature. Each sensor is connected to the MBU via a wireless link and data is sent in regular intervals. The patient's location can be tracked by a GPS (Global Positioning System) or through the position of Internet access point. The MBU conveys data captured from each sensor and existing home WLAN that is linked to the telemedicine system. The electronic patient record can be automatically updated with the received data. In the event of an imminent medical condition being detected, an alert will be generated and patient's location is known so that necessary attention can be provided.

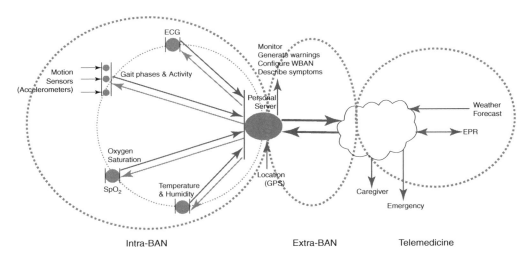

Figure 3.1   Body Area Network connected to the outside world via a telemedicine link

The patient's environmental conditions like ambient temperature and humidity can also be known and recorded. Quantitative analysis of various conditions and patterns for issuing appropriate recommendations can be easily achieved. Data can also be stored anonymously for the purpose of research so that the effects of each parameter on a given medical condition can be analyzed. Legal regulations in most countries may restrict access to patient-identifiable information.

The effects of the human body on propagation characteristics in BAN signal transmission can be an important issue to study since sensors may be facing different directions when placed on a patient (Wang, 2009). When the person moves, some sensors may face nearer to the MBU while others may be moved further away. (Welch, 2002) discussed attenuation and delay induced by the human body as signal degradation factors caused due to absorption, reflection, and diffraction. The electric properties of human tissue (i.e. the electric conductivity and permittivity), can be used to determine the behaviour of radio signal propagating through the human body. Generally, the relative permittivity decreases when the conductivity increases with the increasing signal frequency. A detailed description of these electric properties of the human body on wave propagation is given in (Means, 2001). Measurement, often can be done with appropriate simulating models instead of employing human subjects, is usually necessary to verify the network performance during the design stage to ensure reliability.

3.2 Emergency Rescue

Accidents do happen anytime anywhere regardless of how careful people are. Mishaps can be caused by nature, intentional or unintentional manmade commotion, machinery failure, or a combination of these. In the event of an accident that leads to injury, the utmost priority is always to provide appropriate treatment at the very earliest opportunity. Traditionally, the course of seeking help can be a very lengthy process. Minimizing the time to provide treatment is often the best way to save life, and telemedicine offers the solution for emergency rescue in this respect. How wireless technology can help is very obvious, an example can be as simple as the popularity of cellular phones over the past two decades when people can immediately take a mobile phone from their pocket to call for an ambulance from virtually anywhere, the amount of time saved compared to the era when it was necessary to look for a fixed line telephone; such difference can potentially differentiate between life or death to an injured person. This, of course, is only made possible so long as the cellular phone is within service coverage area thus extending coverage will improve the chance of saving someone. When used in conjunction with GPS, the caller's location can also be automatically reported. Wireless communication and multimedia technologies are combined in many ways for Emergency Medical Services (EMS) as a diverse range of highly mobile devices become available.

Telemedicine can do far more than this. (Ansari, 2006) outlined many deployment options that wireless telemedicine can serve different situations in case of emergency. Cellular phones equipped with cameras can do much more than calling emergency centre for assistance. Amongst various examples (Martinez, 2008) reported the use of cellular phones for remote diagnosis to transmit information about colour change detected which results from exposures to disease markers. In this particular case, test strip images are sent indicating the presence of certain kidney diseases. To serve this purpose, the cellular phone's camera will suffice so long as the 'colour depth' is adequate to distinguish between different colour changes reflecting the properties of the fluid under test. Here, the 'colour depth' is determined by how many binary bits are used to represent each primary colour, namely red, green, and blue, of a given pixel in the image. A camera whose colour depth is n bit is capable of capturing an image of 2^n different levels of shades of each primary colour. Since the indication of the presence or absence of a substance does not require distinction of subtle colour change, an ordinary cellular phone is good enough for such an application. However, in other situations such as capturing images showing a wound, the required image quality may far exceed that of what a cellular phone built-in camera can capture. It is therefore necessary for more sophisticated devices in emergency rescue missions.

There is a huge range of information that telemedicine systems can capture remotely, we examine a case study where Figure 3.2 shows the framework of an emergency rescue system capable of providing paramedics a convenient medium for sending a large amount of

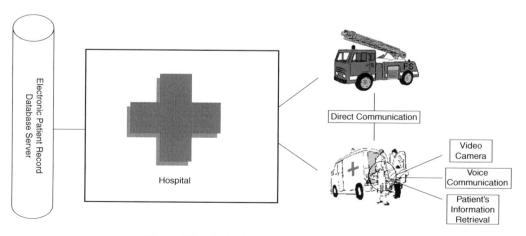

Figure 3.2   A simple emergency rescue system

information about an injured person to the hospital so that necessary preparations can be done prior to the patient's arrival. We look into the details of this system. In situations where fire engines are also required on site, direct communication can be provided for linking paramedics and firemen to facilitate collaborative operation. Each paramedic carries a number of wearable devices, including camera, sensors, and communications equipment. Similarly, a fireman can also wear tracking devices, gas detectors, and oxygen level indicators. The ambulance, which serves as an access point for all paramedics, will provide a two-way communication link between the paramedics and the hospital, where paramedics can retrieve a patient's medical history from the electronic patient record stored in the hospital so that information such as allergy and health conditions can be known when first aid treatment is provided as described below.

3.2.1   At the Scene

A WLAN serves the proximity of the ambulance while attending an accident scene. Its simultaneously connecting devices are carried by several paramedics so that data can be sent towards the hospital. Conversely, information about a patient can also be retrieved from the electronic patient record stored in the hospital's database.

Covering the area surrounding an accident scene, unless the recovery manoeuvre takes place deep within a high rise building or a densely vegetated forest where the ambulance is unable to be parked nearby, a typical IEEE 802.11n WLAN will normally suffice. One major advantage of such a network is that line-of-sight (LOS) is not necessary to maintain a connection.

This network can perform the following: wearable camera captures high resolution images showing details of the injury sustained by the patient, image processing algorithms that estimate the amount of blood loss performed by object extraction that approximates the volume of spilled blood corresponding to the loss, various sensors acquire respective vital signs such as heart and respiratory rate, SpO_2 level, etc.. In providing immediate healing, a paramedic may need to rapidly retrieve the medical history of the injured patient, information such as drug allergy and carriage of any spreadable disease is vitally important so that necessary precautions can be taken. Video conferencing technology also makes consultation with specialists much easier, especially when the patient's conditions are sent to the specialist for providing remote advice.

So, a vast amount of data is captured by different devices of each paramedic. There are various issues that need to be addressed. First, identification of each set of data must be clear in situations where more than one patient is treated on-scene, i.e. to which patient a given set of data belongs must be readily distinguishable. Checking to see whether all paramedics are well within the network's coverage area, as it is necessary to ensure that paramedics can move freely near the ambulance with the assurance that they remain connected at all times. Data security must be addressed to ensure that patient's information will not be stolen by unauthorized personnel nearby, and at the same time not interfered or tampered with during transmission in an uncontrolled environment. To get some idea about how much data is collected by each paramedic, Table 3.1 lists an example of a paramedic carrying the devices described above. This may not seem to be a high demand compared to our daily usage of the Internet, one important difference to remember is that in the event of a failure we can easily reload a page while surfing the Internet, but in life-saving telemedicine applications the circumstances may not permit a second round of data capture and re-transmission should a problem arises. So, adequate resources must be provided to ensure a certain margin for error is accommodated.

Table 3.1 Data requirements

Source	Format	Approximate data rate	Compression
Video Camera	25 fps 1280 × 720	19 Mbps	Yes
Still Image (each)	3000 × 2000, JPEG	1.5 MB	Yes
Voice	3 Hz bandwidth, 32 KHz sampling	525 Kbps	No
ECG monitor	12 leads ECG	12 Kbps	No

In section 2.4 we discussed the issues associated with outdoor wireless communications. In situations where paramedics attending to an accident scene will often find diffraction and reflection the utmost degradation factors to the connectivity of their devices. To see what the possible issues are, we zoom into this portion of Figure 3.2 to show the surrounding of the ambulance in Figure 3.3. As the paramedics are likely to move around during the rescue operation, there is no assurance that the transmitting devices will maintain a clear LOS to the AP's antenna at the ambulance. The amount of diffraction primarily depends on the geometry of the obstructing object, and properties of the signal such as the amplitude, phase, and polarization of the carrier wave at the point of hitting the object. Sometimes the wave may also be partially diffracted during the process of reflection. The extent of reflection and diffraction are primarily dependent on the material of the object it hits, and generally affected by the polarization and the incident angle.

3.2.2 Supporting the Paramedic

A number of wearable devices may be carried by a paramedic depending on the nature of rescue and types of information sought. Figure 3.4 shows a collection of wireless equipment that a paramedic wears when attending to an injured patient. Due to design consideration for small wearable devices, it is often advantageous to set up a BAN using Bluetooth instead of

Figure 3.3 Data communication around the ambulance

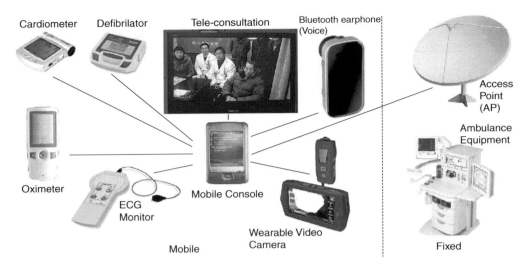

Figure 3.4   Wireless devices serving a paramedic on the scene

connecting an individual device directly to the ambulance as part of the LAN. In this particular example, a customized PDA acts as an MBU that collects data from all sensors and cameras worn and it is connected to the ambulance network via a 2.4 GHz link. (Baber, 2007) discusses situations where wires may sometimes be preferred when hooking up wearable devices, the paramedic's posture and the desired device interface may make wires more preferable than wireless options like Bluetooth or Zigbee. In general, small devices fitted in a pocket that require minimal interaction during usage can be connected via wires. Wires are usually more reliable and data communication will not be affected by movement and orientation, it should therefore be a preferred option in situations where wires will not be tangled and user's movement will not be affected in any way.

No matter what an individual device's function is, the wearing comfort and ease of use must be taken into consideration during design. Power consumption is an important factor to reduce size and weight. Also, reception properties in relation to movement and orientation need to be thoroughly studied for optimal operational reliability. Most of these wearable medical devices are highly customized so very few off-the-shelf apparatus are available on the market. Ergonomical design is a vital attribute to ensure that when worn the device will not affect the paramedic's normal duties in anyway while capturing data. Advances in programmable digital signal processing (DSP) chips enable one single type of processor to be tailored to drive virtually any sensor.

Specific supporting devices worn depend on the circumstances of each rescue mission. For example, illumination may be necessary for night operation and this requirement will draw more power hence battery life may be significantly shortened. Most devices require waterproof housing for reliable operation under heavy rain. Secured mounting ensures that nothing will fall off while the paramedic runs. Many devices are available depending on the type of information required, it is even possible to implement physiological monitoring in disaster recovery where paramedics may be stretched to work non-stop for many hours.

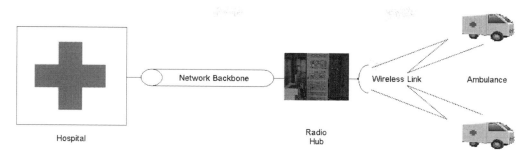

Figure 3.5   Emergency rescue network block diagram

3.2.3   Network Backbone

The network backbone can be virtually any type of wired or wireless network that provides the necessary coverage and bandwidth. In this particular example, we look at the 17 GHz wireless network illustrated in Figure 3.5 (Fong, 2005a), which effectively expands Figure 3.2 to include the details of the part of the wireless network connecting the ambulance to the hospital. This provides a two-way wireless link between the ambulance and a radio hub at the hospital. This is an IEEE 802.16 point-to-point network that would work well when the ambulance remains stationary at the accident scene, but its performance will significantly deteriorate when the ambulance moves. Mobile WiMAX would be a better option for moving vehicles if continuing assessment throughout the course of travelling back to the hospital is necessary. However, since most, if not all, vital information is acquired from the scene mobility support is generally not essential for accident recovery.

Communications between the hospital and the devices featured in Figure 3.6 worn by the paramedic may not be available at all times. For example, tall structures may leave no LOS path within the proximity of the accident scene or along the path between the scene and the hospital. From the knowledge we learned in Chapter 2 there are several issues to consider. The best way to ensure network coverage is by surveying the service areas covered by the hospital to establish a terrain elevation database that essentially consists of a computer elevation map. Its main function is to represent the terrain information to model the effects of buildings and trees in the township on communication with ambulances when serving different areas. Any given location z at a particular (x, y) position of the township is representing the relative altitude of the ground above a fixed reference, such as the rooftop of a high-rise building. A comprehensive database of these terrain points (x, y, z) can be viewed as a grid for the entire coverage area. The terrain database should cover anywhere that the hospital serves so that wherever an ambulance attends will be covered.

This communication link requires high reliability and availability while delay is not normally an important factor. Transmission must be error free but since information about a patient does not have to reach the hospital in real-time, data re-transmission is not an issue. In the event when data is either lost or corrupted, it can be sent again. Re-transmission guarantees successful data reception in the expense of time delay.

Heavy rain is well-known to be one significant contributing factor to serious accidents. Indeed, rain not only increases the risk of an accident, it also affects the performance and

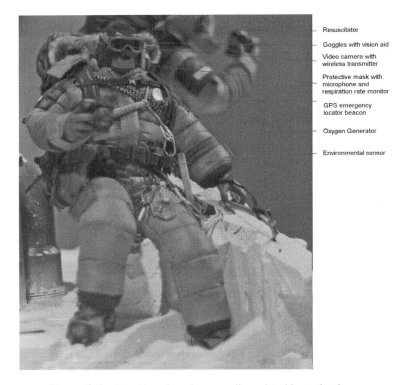

Resuscitator

Goggles with vision aid

Video camera with
wireless transmitter

Protective mask with
microphone and
respiration rate monitor

GPS emergency
locator beacon

Oxygen Generator

Environmental sensor

Figure 3.6   A well-equipped paramedic assisted by technology

reliability of radio links. Most notable problems caused by rain include attenuation and depolarization. The former weakens signal strength that leads to a reduction in coverage, whereas the latter has very substantial impact on wireless links utilizing both vertical and horizontal polarization signals as depolarization may cause the two signals to overlap with each other that eventually end up with no signal at all. To illustrate how severe these problems are, we demonstrated earlier in Figure 2.8 the extent of rain attenuation as a function of variation of rain intensity showing how much heavy rain can affect wireless networks that operate in an outdoor environment. In this plot, we compare the effects of rain on 10 GHz and 5 GHz signals between rainfall rates of 0 mm/hr and 120 mm/hr. From Fig. 2.9, we notice that a horizontally polarized signal is generally more severely affected by rain under identical conditions. Therefore, depolarization will eventually cause the phase shift to undergo as much as 90° resulting in the horizontally polarized signal overlapping with the vertical polarization signal. As the two signals overlap, they combine together and effectively cancel out each other.

The network backbone is a vital part of the telemedicine system that provides a reliable link between personnel working on site and the hospital together. (Fong, 2005b) proposes that generally where licensing of radio frequencies permit, lower frequency of 10 GHz or below should be considered for tropical areas where heavy persistent rainfall of over 20 mm/hr is frequently expected. Otherwise, anywhere between 25 and 40 GHz should be used for optimizing the backbone's performance and to avoid spectrum congestion.

3.2.4   At the Hospital

As information showing the extent of injury a patient suffers in an accident is sent to the hospital along with vital body signs, surgeons in the Accident and Emergency (A&E) unit can get a good idea about what to expect when the ambulance brings the patient in. Electronic patient record can be automatically retrieved so that the patient's medical history can be known. While the benefits to A&E personnel brought by telemedicine technology is clear, there are still a number of challenges that need to be dealt with. Earlier work by (Benger, 2001) has identified a number of potential problems that arise from the expansion of service capability, listed are a number of human factors like convenience, reliability and integrating telemedicine into current practice.

Surgeons and supporting staff need to familiarize themselves with the system, what it delivers and how it can be fully utilized. This may inevitably entail training to ensure that information delivered by the telemedicine system is correctly interpreted. Integration with an existing medical system may also demand special consideration as linkage of a proprietary system may involve compatibility and interoperability issues. Potential cause of interference is a topic that warrants investigation as transmitting devices are used in telemedicine systems although (Tachakra, 2006) has reported that no noticeable interference between the telemedicine transmitting devices and delicate medical instruments in A&E has been detected.

Telemedicine is capable of providing vital information about a patient prior to arrival. Information such as heart and breath, images showing extent of injury, vital signs such as heart and respiratory rates, pulse oxymetry (SaO_2/SpO_2) and arterial blood oxygen tension (PaO_2) levels, diastolic arterial blood pressure (DABP) can all be made available and updated as the patient arrives. Although a wide range of information can be sent, most of these do not incur a large amount of data therefore channel bandwidth is generally not a problem. Some systems also support real-time video conferencing capability that may require a data transmission rate of over 1 MB/s.

3.2.5   The Authority

Since e-health entails patient surveillance, privacy therefore becomes a primary issue for authorities concerned with any possible lawsuit that seeks damages for breach of security. Liability issue is therefore one impediment that may impact the popularity of telemedicine. Deployment crossing state boundaries can cause regulatory issues if service spans different states with different legal and licensing directives. Initial deployment expenditure and lack of funding may also be an important issue limiting exploitation of telemedicine for A&E as the cost benefits may not be too obvious to officials even though the precious time saved in treating an injured patient and ultimately saving a life can be exceedingly significant. Authorities often make decisions based on a business point of view; sometimes the monetary investment is anticipated to yield financial returns within a specified time frame. So, authorities need to be convinced of the perceptible advantages brought by telemedicine.

Technological challenges may not be as difficult to conquer as obtaining government endorsement in many cases. Setting up a comprehensive network for supporting emergency rescue may entail co-operation from various parties discussed in this section. Also, time needed to provide adequate training for healthcare professionals of varying capacities may be perceived as a time consuming process for authorities. The long term benefits brought to life saving are very obvious, yet gaining support on the financial and time investments needed is another issue that needs to be worked on.

3.3   Remote Recovery

Wireless telemedicine facilitates healing just about anywhere, on land, at sea, as well as in the air. Over a decade ago, commercial airliners began linking their planes to *MedLink* described in (Mchugh, 1997), as a service to complete healthcare coverage beyond the earth's surface, offering basic life support information to trained airline personnel to carry out basic medical emergency procedures and diagnosis on whether urgent stop is required for medical attention. The underlying technology allows the airline industry to save considerable time and money which an unscheduled stopover to drop off passengers who may not be in need of immediate medical attention.

Through video conferencing, experts in different countries can offer real-time medical advice to service personnel who may not even have any prior healthcare training. It is only a matter of offering recommendations as to what to do. Telemedicine also enable an electronic patient record of a specific passenger to be retrieved so that any existing medical conditions can be known. In addition to assisting patients in the air, telemedicine makes recovery and healing available virtually anywhere. Remote recovery often involves swift discovery of the patient's exact whereabouts, and any potential hazards to the rescuer from the patient can be ascertained to avoid putting them in danger. Technology allows telemedicine to assist remote recovery in these situations. We shall look at three situations where telemedicine frequently helps to save lives of both the general public and professionals who risk their lives to rescue them.

3.3.1   At Sea

Maritime recovery is a challenging scenario since cellular communication networks cannot serve areas in the sea. In vast oceans where no land can be seen, communication is limited to satellite links. Although most modern vessels are equipped with high precision GPS, this may not always work because the person who requires urgent rescue may be thrown overboard, or the vessel may be sunk or have lost power. Technology makes recovery in many situations much easier than before. Finding a person is perhaps the very first thing rescuers need to do at sea. Video extraction technology used in conjunction with high-resolution video capturing makes locating a person or an object in the sea much easier. Co-ordination between rescue boats, helicopters, and control centre must be supported in real-time as current drift can very quickly move the person to be rescued.

Satellite communication is often used in maritime rescue since this is the only means of providing comprehensive coverage across vast oceans. Since satellite communication utilizes millimetre wave in the GHz magnitude, retrieval of sunken vessels is virtually impossible due to absorption through water at these frequencies. Underwater wireless communication is far more challenging than communications through air. In contrast, acoustic waves propagate some five times faster in water than in air making an acoustic pressure channel suitable for underwater communication. Long-range underwater acoustic propagation study commenced as far back as the fifteenth century. As its name suggests, acoustic channel involves audible frequency range that spans between tens of hertz up to around 20 KHz. So, wireless communication systems have very different requirements in terms of transceiver structure and antennas. As acoustic wave propagates far slower than millimetre wave through air, long transmission delay is expected. In addition to significant propagation delay, underwater communication also suffers from high variation of multipath and narrow available bandwidth.

3.3.2   Forests and Mountains

Search and rescue in densely vegetated areas often cannot be performed visually. Although infra-red cameras can help pinpoint the location of a person in some situations, it is by no means an all-round solution and its effectiveness is limited by many circumstantial factors. Rescue is made even more difficult since cellular phone coverage is highly unlikely to be available in remote forests and mountains.

Radio communication is extremely difficult in these areas. First, it makes no economic sense for operators to provide coverage to these areas given the enormously low subscriber density and utilization rate. LOS link cannot be maintained due to dense vegetation, therefore diffraction and reflection can be important degradation factors that affect successful communication. Remember, the basic mission of a radio link is to deliver sufficient signal power to the receiver so that some kind of meaningful information, such as the receiver's whereabouts or images showing the surrounding area, can be realized. The effects of physical obstacles like plants on wave propagation go back to the concept of clearing Fresnel zone as outlined in section 2.4, which refers to the volume of space enclosed by an ellipsoid of the two antennas between the ends of a radio link. The radio link can be maintained if no objects are within the area to cause significant diffraction into the corresponding ellipsoid. Having said that, it does not necessarily imply that failure to clear the Fresnel zone will always result in loss of communication. The actual network degradation experienced very much depends on the operating environment. The ground reflection path will sometimes be obstructed by the disturbance of trees and other plants whereas ground reflections can be a major factor of path loss in plateaus of squat vegetation or lakes. Although existence of LOS path may not be likely, if it does have some gaps between the transmitter and the receiver, one direct path and a ground-reflected path may both exist. In such a case, the path loss would depend on the relative amplitude and phase relationship of the signals propagated through the two paths. The amplitude and phase of the reflected wave depend on a number of variables, including conductivity and permittivity of the reflecting surface, frequency, angle of incidence, and polarization. They will overlap each other and the effects can vary. The relative signal strength between the two paths would depend on the ratio between the ground-reflected paths having Fresnel clearance and the LOS signal path if present. If the former undergoes little loss due to reflection, the two paths would have similar signal strengths. This situation may result in either a boost of up to 6 dB over the signal across the direct path alone, or cancellation resulting in additional path loss of 20 dB or more depending on the relative phase shift of the two paths. Two signals combined in phase

(without relative phase shift) would result in 'constructive interference' or out of phase (180°
relative phase shift) causes 'destructive interference' as learned in high school physics. Spread
spectrum techniques and antenna diversity are often considered to be effective solutions to
control this problem. In addition, attenuation from fog can be significant at frequencies above
20 GHz and this can be an important consideration for communication systems in humid
forests.

Clutter, defined in *Wiki* as 'excessive physical disorder', is a term often seen in wireless
communications that refer to vegetation that affects signal propagation. Clutter usually causes
attenuation and scattering when radio waves hit a surface resulting from variation of multipath
due to movement of branches and leaves by wind. The extent of scattering usually depends
on density of leaves, leaf shape and the amount of water held within a leaf. It is therefore
extremely difficult to predict the propagation characteristics through forests.

3.3.3   Buildings on Fire

Amongst the types of awkward rescue operations discussed in this section, people recovery
from a fire inferno is most likely the most challenging situation given the amount of time
available to rescuers. Fire can spread very quickly and is almost always accompanied by thick
smoke impairing vision. The combined effect makes finding the exit path very difficult at
times. Paramedics and firemen alike require extremely reliable communication systems and
tools that can bring them back to safety in minimal time. The fact that people without special
needs may not carry any transmitting identification devices makes finding people trapped
more difficult in the event of a fire. It is therefore an unrealistic expectation to locate a missing
person by using a radio because you are assumingthat that the person sought wears some kind
of radio transmitting device that is fully functional. This problem implies that safe recovery can
only rely on professionals risking their own lives to ensure any missing person is found at the
earliest opportunity and to lead the person through a safe escape route to safety. Unfortunately,
a floor plan may not always be available for rescue professionals upon entering the building.
A building on fire can easily turns itself into a maze. Further, the path which they take when
entering the building may not necessarily be the shortest and safest to take for escaping. The
entry path may also risk subsequent blockage by falling obstacles. Having said all this, how
can technology assist them in path finding?

Thick smoke can severely impair vision making the surroundings virtually impossible to see.
Likewise, radio links can be blocked by partitions that comprise energy absorbing materials.
Metal is a particularly 'unfriendly' material for radio waves to get through and it is inevitably
used in buildings for a variety of reasons. Without being able to personally experience such a
situation where vision is blinded by smoke and communication is cut off intermittently when
moving around inside an inferno, it is difficult to describe in words how desperate the situation
for rescuers actually is. Technology is here to assist their operation and to maximize the chance
of a successful rescue. This is an area where telecommunications, in particular, can help save
lives.

We learned in Chapter 2 that a signal of a frequency in excess of several GHz generally
penetrates through materials better than lower frequency waves. We may have experienced
cellular phone service interruption when entering a lift (or elevator), as most lifts are made
of steel enclosures that effectively act as a metal cage that 'shields' off the commonly used

900 MHz signal for cellular communications. For this reason, firefighters need something more reliable than a mobile phone to ensure that they can find the safest exit route. A comprehensively equipped firefighter is depicted in Figure 3.6. The figure shows a rescuer with different apparatus to that shown in Figure 3.4 and also some with data transmission and reception capabilities. Each communicating device has its own function in ensuring the user's safety. The point worth noting is that the rescuer featured wears a protective suit that keeps both the rescuer and the equipment worn well-protected against prolonged exposure to excessive heat and toxic gases. Any communication equipment must be designed to remain connected under any shield within the suit. Further, effective filtering must be in place to ensure that any ambient noise that may affect communications will be removed in order to retain sustainable communications inside a building on fire.

Another vital survival tool is the amount of remaining oxygen needed to ensure that breathing can be sustained until reaching a safe location. Advanced alert must be generated to allow adequate time for escaping and in the case of any mishap a rescue team can bring in an additional oxygen supply before exhaustion. In the process of issuing such an alert, it must be made in a subtle way to avoid putting unnecessary pressure on the firefighter to ease any additional anxiety. In addition to oxygen supply status, detection of any flammable or toxic gas and, if available, video footage showing the site's environment, can be reported to an off-site control centre or command post in order to build a better picture about what is going on inside a blaze. So, a reliable network that supports a range of communication needs is necessary to keep rescuers safe. A report by (TriData Corp., 2005) pointed out a number of deficiencies with the conventional VHF (Very High Frequency) radio of 30–300 MHz range used by US fire departments. More recently, FCC assigns the 800 MHz band for public safety radio communication in an effort to reduce spectrum congestion that spans across the range used by commercial broadcast and effect of interference. It is reported that different radio channels are often used within a fire department. Interoperability is therefore said to be a major issue. So, is it really feasible to standardize firefighters' communication systems?

On hindsight satellite may sound good due to its vast coverage and excellent penetration properties. Satellite phone may be a good choice for the sole purpose of providing a medium for talking to off-site supporting personnel. However, the precision for location tracking is far from being adequate for fire rescue within a building. The precision of GPS positional accuracy depends on uncontrollable factors including satellite placement and the effective DOP (Dilution of Precision) that can be severely by nearby buildings. Normally, GPS can only identify a location within a radius of several metres at best. This may mean that someone in an emergency may be mistakenly identified as being trapped in an adjacent room and this consequently leads to increase in seek time. Such precision deficiency may even lead to a search being conducted on the wrong floor of the building where 3-D positioning is not used. Satellite is therefore not a suitable solution for fire rescue. Other solutions such as RFID for short range path recognition and marking can be explored, since markers can be placed automatically along the entry route during an operation.

There are several fundamental requirements that need to be satisfied: light weight and easy to operate with minimal intervention, precise position tracking, good penetration through various materials commonly used in building construction, and resilience to excessive heat. Until now there is no single technology that satisfies all these requirements. The most likely viable solution is therefore to integrate different solutions with a high degree of interoperability kept in mind.

In this section, we have looked at three different demanding situations for rescue operations. Each of these has different fundamental requirements and problems. The only thing they all share is an exceptional level of dependability and ease of operation. Through technological advancements of different types of communication networks, more sophisticated wireless systems can be developed to meet the growing needs in an effort to improve the chance of survival in difficult situations.

3.4   At the Hospital

Information technology has brought automation and safety into hospitals over several decades. The list of improvements IT has made to the way a hospital runs is endless (Felt-Lisk, 2006) has given an example of how six different areas of IT make a significant difference to a hospital. Due to the vast possibilities of IT applications in healthcare it is simply impossible to cover everything in one single book volume, therefore, we shall concentrate our discussion on how communication technologies help modernize a hospital by first briefly reviewing a case documented in a magazine article (Mullaney, 2006) entitled, 'The Digital Hospital', followed by going deeper into how telemedicine plays a momentous role in the daily operations throughout different departments of a hospital. The article started by reporting a case where a physician received an automated warning when he requested a drug to be dispensed. An information system in the hospital has detected a possible risk of mixing this particular drug with one that the patient has previously taken. Such an alert prompts the physician to prescribe an alternative medicine as a remedy to eliminate the detected risk. This is just one of the many examples where the timely delivery of information easily saves lives. The article then went on to discuss physicians examining X-ray images and controlling a robot inside the hospital that can be accomplished remotely. All these, plus many other tasks are made possible through telemedicine.

A hospital that provides comprehensive services may be composed of many departments with a central administration. Figure 3.7 shows a simplified version of a typical hospital

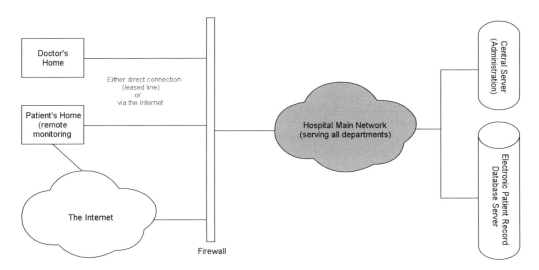

Figure 3.7   Block diagram of a typical hospital network

having several departments all housed in a single complex, all linked together by a network for information sharing and co-ordination. Obviously, each department would have its own requirements on the type of information processed and the urgency of information retrieval prioritized. For example, A&E for treating an injured patient would require information about the patient much more urgently than Paediatrics providing a general health assessment. Both of these involve retrieval of medical history and update of new information. The tolerance to delay and the ease of reading retrieved information are very different for these two examples. In section 3.2, we have seen the importance of telemedicine on maximizing the efficiency of the A&E department by providing surgeons with the necessary information about an injured patient even prior to arrival. We shall look at three other examples where communication technology can help save costs and lives in a hospital. Since there are so many different situations where telemedicine finds its importance at a hospital, the examples below are:

- Cost saving measures in radiology with accurate and timely information.
- Precision control of robots for surgery.
- Reliable tracking of newborn babies to ensure misidentification never happens.

These examples are selected to illustrate different categories of wireless communication applications, namely quality assurance, remote sensing, and surveillance. Many other situations can utilize telemedicine with very much the same underlying technologies.

3.4.1   Radiology Detects Cancer and Abnormality

Radiology is an important area of medicine for early diagnosis and treatment to ensure maximum chance of survival. This is an application where communication is critical both among hospital staff and patients. So, telemedicine extends beyond the hospital network. Delay in the delivery of correct information to the appropriate parties may lead to unnecessary delay of treatment and this may result in legal consequences. Radiology involves accurate interpretation of medical images. The images by themselves frequently do not make any sense to the patients so contextual explanation is an important part of communication between the hospital and the patient. Images are therefore accompanied by reports. Figure 3.8 shows a block diagram

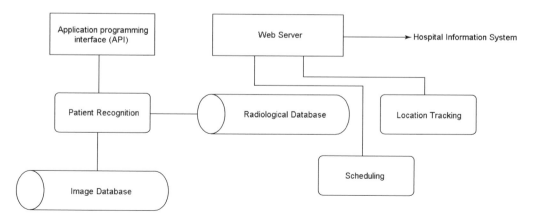

Figure 3.8   Case study: radiology information system

of what the radiology information system entails. Permission must be granted individually to hospital personnel involved to avoid any risk of unauthorized access and undesired alteration of information. In this system, an SMS (Short Message Service) text message is automatically sent to the patient as a reminder for a check up through the scheduling module. When the patient arrives at the hospital, an RFID card informs the Radiology Department staff members of the patient's arrival and the electronic patient record is automatically retrieved and it will also be used for tracking. Results are sent to the hospital information system for analysis so that any necessary actions can be taken. Archives will be stored in separate databases for images and radiological data.

One major objective of effective communication is cost saving, ensuring the proper delivery of information can potentially save a large amount of money as (Brenner, 2005) reported the cost of an erroneous communication averages over US$ 200 000 per case. So, the entire process shown in Figure 3.8 from the radiologist capturing the image to delivering an extract of its associated report to the patient must be made error-free to ensure proper communication of information is attained so that cases of indemnity payout are kept to an absolute minimum. This is best achieved by proper design and maintenance of a related telemedicine system. Referring back to Figure 3.8, a radiograph is taken with a patient in the X-ray facility location with one radiologist serving a number of patients per session. These images are captured, digitized and sent to specialists handling respective patients.

Next, we investigate what can possibly go wrong during the process. The worst possible calamity that could ever happen is mixing up images of different patients which leads to incorrect diagnosis of a healthy person with cancer while the cancer sufferer is wrongfully discharged. Obviously, such misdiagnosis will lead to negative psychological consequences to patients and their families, and a healthy patient being unnecessarily operated on while leaving the other patient with cancer undetected. The first line of safeguarding each image and ensuring that they are correctly referred to the respective patient is by proper filing of each image throughout the entire process. With apposite procedures in place and strictly adhered to, information management can help ensure images are well taken care of. Next, when each image is successfully passed on to the specialist, the image is examined and any abnormal signs detected, either manually by the specialist or with automated feature extraction using image processing techniques. At an early stage of tumour formation, especially in the CIS (Carcinoma in Situ) stage, subtle signs may not be easily visualized. This is a critical time to prevent the invasive phase so that more treatment options are available. Noise free high resolution images without loss of fine detail is therefore vitally important in the image transmission process. Radiographic images are usually transmitted digitally and image clarity very much depends on the 'bit error rate' (BER), that effectively measures how many bits are sent when one bit is corrupted in the data stream. Ultimately, we never want any subtle sign showing the tumour to be missed out due to a few bits missing from the digital image.

3.4.2 Robot Assisted Telesurgery

The term 'telesurgery' refers to a surgical operation being carried out by surgeons remotely without physically being in the operating theatre. High precision robots have only recentlybeen made possible with the availability of tiny sensors and actuators. These small actuators make very small movements that usually involve moving in all three dimensions. The primary

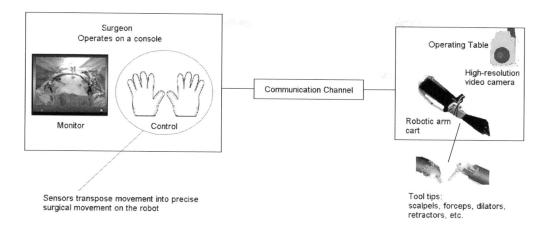

Figure 3.9   Tele-robotic surgery

function of an actuator is to initiate a robot's movement based on an instruction given by the surgeon. Figure 3.9 shows how a surgeon can carry out an operation remotely by using a telemedicine system to control a robot in the operating theatre. In addition to hand controls, (Randerson, 2008) reports that eye-controlled robots make 3-D mappings of tissue possible and automatically calculate the depth of tissue by tracking the controlling surgeon's eye movement to precisely track the area where the surgeon is operating.

Robotic telesurgery effectively brings a surgeon's professional techniques into an operating theatre that does not have a surgeon physically present. However, in order to make this happen a large amount of data exchange is involved between the surgeon and the robot 'acting on their behalf'. For a start, the surgeon needs a good view of what is going on inside the operating theatre. Cameras are installed in the operating theatre and they must incorporate remotely controllable rotation and high-power zooming functions. Also, the video image captured must be displayed next to the surgeon in real-time without any noticeable delay so that any movement of the robot will not be delayed. Even a very small amount of time delay in the robot's action can lead to irreparable damage to the patient's body. Time delay (latency) is a big issue with long distance telesurgery. However, with telecommunications that span across continents transmission delay is an unavoidable issue. This is likely to be one of the most challenging issues for long distance operation.

User interface for control must be carefully designed to ensure that the entire system co-operates well with the surgeon, voice activated control would ensure minimal disruption is caused during the operation. This involves speech recognition algorithm that not only correctly interprets each individual command issued by the surgeon, but also identifies the voice of each individual person within the vicinity so that only the respective surgeon's command is acted upon. This is vital in ensuring that voice commands from other surgeons or supporting staff will not be mixed up and inappropriately acted upon. Robot control requires exceptionally high precision 3-D hand movement manipulation, commonly through a pair of virtual gloves. Sometimes the term 'six-dimensional' is used to describe the sensors in these gloves. The 'six dimensions' are just the positive and negative directions along any of the 'x', 'y', and 'z'

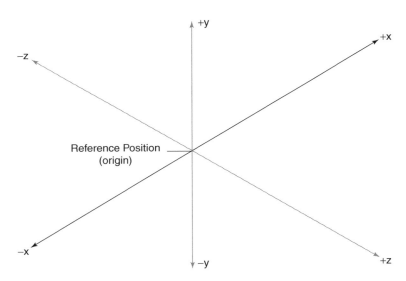

Figure 3.10   'Six-dimensions' representing the 3-D space

axis representing the 3-D space away from any fixed reference point, as depicted in Figure 3.10. The sensors' movements drive the respective actuators in order to control the robotic hand that operates a piece of surgical tool, including the change of different tools on the robotic hand. In addition to control signals, a voice channel should also be given for video conferencing between the surgeon and personnel inside the operating theatre. So, telesurgery involves transmission of high resolution real-time video images and control signals of high precision. Minimizing time delay is a vitally important issue for successful implementation of robot assisted surgery.

Portable robotic surgeons would be extremely useful in demanding remote rescue missions such as in the examples discussed in section 3.3. In the worst case, only an expensive robot will be written off without putting any precious life into jeopardy. They can even go underwater if necessary (Blackwell, 2006). There are in fact many situations where robots go into dangerous situations for high risk rescues.

3.4.3   People Tracking

We commence our discussion by emphasizing the need to keep track of babies in a hospital as an example of how technology can help prevent mistaking individual babies among a group. The word 'tracking' in our example has nothing to do with any possible breach of privacy that involves surveillance. Cases of newborn mix up have been reported throughout the world and often leads to avoidable yet substantial emotional damage payouts. Most mix up cases are indeed avoidable since they are direct consequences of irresponsible personnel failing to follow all necessary procedures for handling babies. The good news is that foolproof technology is here to help eliminate such risk simply with a couple of tags that cost a few pennies (a mere

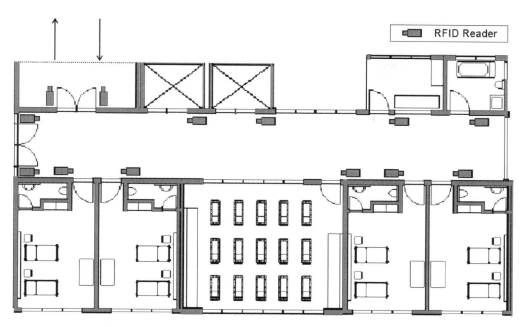

Figure 3.11   RFID readers installed in a hospital maternity ward

10 US cents). By referring back to section 2.5 where we talked about RFID, it is not difficult to understand how RFID tags can help identify each individual baby.

Figure 3.11 shows the layout of a typical hospital maternity ward where RFID readers are installed on both sides of the main entrance and each nurse carries a handheld reader. When a baby is born two RFID tags are attached with separate submersible bands. The band must be comfortable whilst not too loose to pose any risk of falling off. The reason why two tags are used instead of one is solely for redundancy, so that in the event of any unforeseen problem there will still be another one for identification or confirmation. Since RFID tags are small and light, they are highly unlikely to either cause any discomfort to the baby or inconvenience to either hospital staff or parents handling them. Also, RFID tags are fairly robust and they can be submerged in water so they need not be removed when bathing the baby. A quick scan of the tag can affirmatively identify each baby even though they may look very similar to each other.

It is even possible to track the movement of each baby if active (battery operated) tags are attached so that every time the baby (and the tag) passes a reader that is installed in a given location there is a record of the whereabouts of the baby. This also prevents unauthorized carriage of the baby away from a certain area simply by triggering an alarm when the tag comes close to the exit. However, this alarm system does not prevent tag removal unless the band is tough to make cutting difficult. So, to make it more secure additional circuitry can be included to activate an alarm when the tag is tampered with or when the tag remains stationary for a specified time, say one minute, a long enough time to reasonably assume the tag has been removed or else the baby's movement would confirm the tag remains intact. This works particularly well with babies because they tend to move frequently even when asleep. The use of active tags may cost more but the benefits they bring are obvious. Typically, a baby only stays in the wards for a few days before going home. For this reason, battery life is not a concern, since even a very small embedded battery can easily power an active tag for over a couple of weeks continuously. Further, RFID usage poses no risk of excessive radiation even to a delicate newborn since the intensity of electromagnetic radiation emitted is even lower than surrounding cellular phones that people carry.

A communication system is necessary for linking the alarm system and pagers carried by hospital staff together so that they can be automatically alerted. This can be easily set up with a console that stores information about all tags issued with a map of reader locations. Information about the type of alert together with location of the nearest reader that picks up the RFID signal can be broadcasted to all staff pagers for necessary follow up actions.

3.4.4   Electromagnetic Interference on Medical Instrument

From the examples above we notice that wireless telemedicine is a vital part of an efficient and reliable hospital, but what about possible EMI that may affect that operation of delicate medical instruments? Radio transmitting devices such as cellular phones may cause malfunctioning of medical equipment and in some cases even critical life-supporting apparatus can be badly affected. As ambient environmental electromagnetic noise from various sources both within and outside the hospital site cannot be controlled, proper shielding of medical instruments therefore becomes the most effective way of ensuring reliability irrespective of interference noise level surrounding the area of operation. However, many instruments in themselves are sources of EMI. For example, cardiopulmonary resuscitation (CPR) draws a vast amount of electrical current that would generate an excessive amount of noise. Proper design of housing with appropriate metal shielding will be able to protect an instrument from external interference.

Operating theatres and ICUs (intensive care unit) are most vulnerable to EMI due to the inherent nature of instruments used. In these areas, it would be necessary to restrict the use of transmitting devices including cellular phones by anyone in the proximity. As for the venue itself, it is possible to set up the partition with an absorption chamber that installs pyramidal polyurethane foam inside the wall. This is a layer of foam that effectively blocks off electromagnetic radiation from entering the venue.

3.5   General Health Assessments

Telemedicine for healthcare extends beyond medical protection for patients in need of special attention. It can also facilitate the general public in maintaining good health in many situations, indoor or outdoor, at rest or on the move. No matter where we are technology is always helping us to optimize our well-being. Information technology is found in many areas of health assessments in daily life. For example, dietary monitoring for those who are concerned about weight gain, colour matching for skin care, calculating the amount of calories burnt during a workout, nutrition intake of a child, baby monitoring alarm, automated reminder for a dental check up, etc. There is always something for all ages. Strictly speaking, even ergonomic design factors of appliances that can affect our well-being due to usage can have a close relationship with IT and healthcare very simply because proper product design eliminates the risk of causing users to require medical attention.

Telemedicine finds its use in many situations, for example, assisting with reduction of obesity for subscribers to weight-loss programs. They can have their body weight automatically sent to the control centre for record keeping and progress tracking. Before ending this chapter, we shall look at a number of situations where telemedicine helps us in our daily life. We take the availability of such technology for granted when using it on a regular basis, let us look at how they work in some examples.

Case Study 1: Fitness Monitoring for a Morning Jog

Since off-the-shelf foot-contact pedometers can only count the number of steps taken, accelerometers and gyroscopes are often used for motion monitoring. (Bouten, 1997) reports that a sampling rate of around 18 Hz is adequate for sampling human activities. (Pappas, 2004) and (Bamberg, 2008) have conducted comprehensive studies by installing Shoe Integrated Gait Sensors into running shoe insoles as illustrated in Figure 3.12 with combinations of accelerometers, gyroscopes, electric field sensors, piezoelectric sensors, and resistive band sensors (Morris, 2002). This set of sensors is installed to capture foot movement. A tiny transmitter can send the data out for analysis of the level of activity and track the state of the user. This mechanism can also track uneven wear of the heel part of the sole and detect abnormal

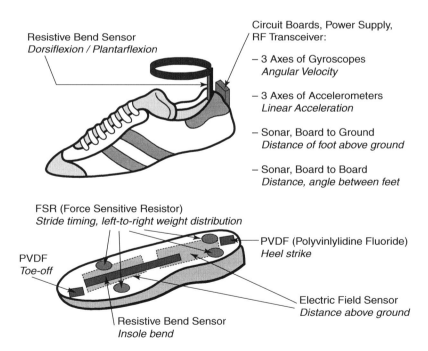

Figure 3.12   Schematic of the shoe-integrated gait sensors. Reproduced with permission from Morris and Paradiso © 2002 IEEE

wear patterns for providing a remedy for running comfort. Technology can help us keep track of how far and how fast we run, as well as reducing uneven wear of our shoes.

Some people may jog with the intention of losing weight, this is also an area where telemedicine helps. Exercise accelerates digestion so one may feel hungry after running. Communication technology can help us activate a microwave oven so that it prepares our breakfast, at a certain predetermined stage of the jog, say the last one kilometre before returning home, a signal can be automatically sent to the smart home control console to start reheating the breakfast so that it is ready by the time the jogger returns home. This is just one simple task smart home automation can accomplish. A coffee brewer can also be activated in very much the same way. After either pre-programming in advance or remotely through a cellular phone, no further action is necessary unless manual override is desired.

A morning jog not only strengthens our muscles, it is also a gentle cardiovascular workout that optimizes both respiration and blood circulation. It helps ease any digestive problems that may accumulate due to busy work schedules. While we can feel the benefits ourselves, technology lets us quantitatively realize the difference and keeps a record of our progress for us by logging our daily activities such as length of jogging route, number of steps taken, duration, heart and respiratory rates. A wearable pulse meter can be bought cheaply and some can even be part of a wristwatch, this little device helps keep track of our heartbeats while jogging. Technology can also help us monitor what we eat and automatically generate a report of nutritional information of each meal throughout the day. So, the after-jog breakfast can be prepared according to the amount of calories burnt. By linking the health monitoring devices to a home PC, the user can check the improvement of health on a daily basis and retrieve a meal recommendation list that is generated from the captured data for optimal nutrition balance.

Although running in an outdoor environment in fine weather may be more pleasurable than in a gym, sometimes gym workout is more desirable as different types of equipment offer full body exercising and it is weather proof. So, this brings us to the next case study of gymnasium health monitoring.

Case Study II: Gym Workout

Many gyms offer free WiFi Internet access even though a physical workout does not normally need it. We may not surf the Internet in the gym but a wireless network does offer a range of possibilities for keeping track of our activities there. Just as we could wear for a morning jog, small sensors can be worn across the body for reading different signs depending on the nature of exercise. Walking or running on treadmills or steppers apply the same technology as that used for morning jogging, almost identical devices except that downloading of captured data is much easier in this case as the gym wireless network can readily support continual downloading of data so that no memory storage is needed within the BAN of sensors and related electronics. Figure 3.13 shows the block diagram of a gym equipped with fitness equipment most commonly found, including treadmill, stepper, weights rack, leverage bench press, elliptical trainers, exercise bikes, and seated rowing machine. Although there are many types of apparatus, how technologies facilitate health assessment can be very similar if we group them according to the way they are used. For example, in the health assessment perspective weights rack and leverage bench press are similar in nature as they both involve using the upper part of the body to lift a certain amount of weights. Weight-lifting is intended for muscle building. The result is best judged by the growth of muscle that can be detected by examining the change of body

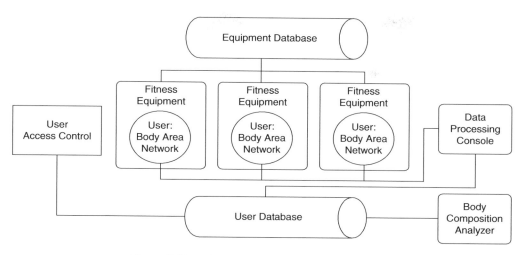

Figure 3.13　Block diagram of a gymnasium network

shape as progress is made. A quick scan of the appropriate part of the user's body can be sent for storage and enables subsequent comparison of body shape change when the user returns in the next session. Technology can help beginners by offering guidance on the proper way of handling dumbbells so as to avoid injury. This can be done by projecting an image to illustrate the correct procedures so that the user can follow it step by step.

Multiple users can be identified in many ways. The most convenient methods are either an embedded passive tag on the user subscription card or placing an RFID sticker on the shoe with readers on the mat associated with each piece of equipment. How it works is very simple, once a user steps on the mat the unique identification number will be recorded, when the equipment is started readings will be captured and marked as belonging to the identified person. In addition to serving the purpose of health tracking, the system can also be used for billing purposes if usage is charged on a per use basis. Upon completion of a session, a user can choose either to download the data onto a removable storage device to bring home or to have it sent home via the gym network. Proper user identification procedure would ensure data associated with each individual user will not be mixed up and privacy is assured.

Case Study III: **Swimming**
Underwater wireless communication always posses difficult challenges as we explained in section 3.3.1 when we discussed the difficulties with soaking a transmitting device in water. Despite the challenges, wireless communications can help save life above and beyond health assessment capabilities similar to that in a gym environment. This is particularly advantageous in beaches where lifeguards may not be able to keep an eye on all swimmers. Any small waterproof transmitter can be used to call for help in the event of an accident. Warning can also be issued from land in a sudden emergency situation such as the citing of a shark, all it takes is a waterproof receiver that picks up broadcast signals from the shore.

So, a system that allows small transceivers to be brought with a swimmer can potentially save lives. Since a swimmer does not go far below the water surface and the distance away from the shore does not normally exceed a couple of hundred metres, water absorption does not necessary block off radio waves completely. One important issue to bear in mind is the material used in waterproof housing since this will also have an effect on signal absorption. Another fact worth noting is that wave penetration properties differ between salt water and with traces of bleach in a swimming pool. Since the data rate generally does not exceed one kilobyte per second, an Underwater Wireless Acoustic Network (UWAN) would do the trick. The major drawback of such a network is severe propagation delay that can be as much as one second per kilometre. This must be taken into consideration during the system design stage.

Setting up a UWAN can be quite complex, swimmers move around inside water hence it is highly unlikely that the transceiver will remain stationary. To illustrate the effect of changing speed we shall look at some basic mathematics here.

Given that the combined speed of swimming and water flow v is at an angle relative to the acoustic signal propagation direction θ, the effective acoustic propagation speed v' is:

$$v' = v.\cos\theta \tag{3.1}$$

Logically, the effective propagation speed v' increases if the combined speed v is moving towards the same direction as signal propagation, whereas v' decreases when v moves in the opposite direction of propagation. The water flow will result in a slight bending of a narrow acoustic beam in the same direction, but its effect is reasonably insignificant. The propagation speed changes significantly when entering a different medium, namely from water into air or vice versa. This effect is due to refraction as the dielectric constant changes just as light bending from air through water or glass. So, refraction will change the direction of the propagating signal. Note, incidentally, that the term 'refraction' is also used in optometry that refers to the examination of an eye in the process of evaluating whether a spectacle prescription enhances vision. Such application is sometimes known as *refractometry*.

In addition to refraction, reflection will also occur when the signal hits the boundary between two media, resulting in a portion of signal being reflected back into water from the surface without going into the air, as shown in Figure 3.14. In shallow waters, as in the case with most beaches and swimming pools, reflection from the bottom will also induce multipath effect.

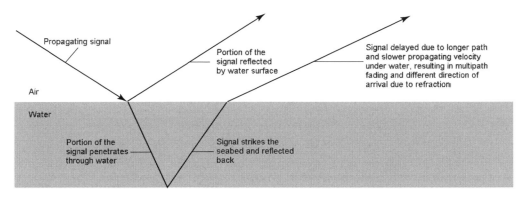

Figure 3.14   Water surface causes reflection and refraction

The received signal $r(t)$ can therefore be expressed mathematically as:

$$r(t) = \sum_{n=1}^{N} \alpha_n.s(t + \tau_n)$$

(3.2)

Where the attenuation coefficient α_n denotes the reduction in signal strength due to absorption attenuation that effectively turns the signal energy into heat and loss due to reflection, which is both frequency and distance dependent; and the original transmitted signal $s(t)$ is subject to a delay of τ_n resulting in $s(t + \tau_n)$. N is the number of incident acoustic signal paths caused by multipath effect. In shallow depth with short range, it will likely be $n = 3$ since there are three signal paths: direct LOS between the transmitter and receiver, a reflection from the surface and another from the bottom. n and τ_n generally increases when the depth and range increases as more reflections will occur and the time for the signal to reach the receiver increases. The reflection loss due to water surface and the bottom can be very different since the bottom may have deposits that make it far from even. The molecular movement of the water surface caused by the propagating signal is very small (the carrier wave is highly unlikely to carry sufficient energy to cause significant movement to water) therefore only a very tiny fraction of the signal will be transmitted from water into air. Virtually the entire signal will be reflected back into the water. Also, acoustic pressure does not couple well with air just like an 'impendence mismatch' with an electrical current hitting a load. A similar situation applies from air into water, this is exactly the reason why when we soak our heads in the swimming pool we can hardly hear anything from above. This coupling problem does not generally exist with the bottom since deposited particles are 'more friendly' with water molecules movement. With better coupling, a certain portion will be reflected back into the water while some will be absorbed. This is good news to communication since absorption will have a negative effect on multipath; the bottom effectively acts as a cushion that shields off some reflected signals thereby reducing n. The actual effectiveness will depend on the composition of the deposit.

Up until now we have looked at signal propagation relative to time. Before we end our discussion lets turn our attention briefly to the effects with respect to distance. Consider the signal $S(d)$ where d is the distance travelled. Obviously, the signal S weakens as d increases. Their relationships can be expressed in basic mathematics as:

$$S(d) = S(d = 0).e^{-\alpha d}$$

(3.3)

Since attenuation is usually expressed in dB, we can represent the signal loss L (not to be confused with the notation in Equation 2.2 that denotes the number of levels there) as:

$$L = 20.\log_{10}\left(\frac{S(0)}{S(d)}\right) = 20.\log_{10}\left(\frac{S(0)}{S(0).e^{-\alpha d}}\right) = 20.\log_{10}\left(e^{\alpha d}\right)$$

(3.4)

This can be simplified as:

$$L = 20.\log_{10}(e)[\alpha d] = 20.[0.434][\alpha d] = (8.86\alpha).d$$

(3.5)

The above discussion gives an insight into the complicated situation of applying telemedicine to healthcare in an underwater environment. Readers are advised to refer to (Etter, 2003) for details on underwater wireless communications.

In this chapter, we have looked at a number of situations where telemedicine can help save lives, it can also be used in applications for general health monitoring so its benefits extend to healthy people too. Wireless communication systems can face difficult challenges in some harsh environments. Barriers such as water and vegetation can significantly affect system reliability, therefore they are not 100% problem free even though technological advancements have made them far more capable than ever before.

References

Ansari, N., Fong, B., and Zhang, Y. T. (2006), Wireless technology advances and challenges for telemedicine, *IEEE Communications Magazine*, 44(4):39–40.

Baber, C. (2007), Can wearable be wireable?, *IET Seminar on Antennas and Propagation for Body-Centric Wireless Communications*.

Bamberg, S. J. M., Benbasat, A. Y., Scarborough, D. M., Krebs, D. E., Paradiso, J. A. (2008), Gait analysis using a shoe-integrated wireless sensor system, *IEEE Transactions on Information Technology in Biomedicine*, 12(4):413–423.

Benger, J. (2001), A review of telemedicine in accident and emergency: the story so far, *Journal of Accident and Emergency Medicine*, 17(3):157–164.

Blackwell, A. (2006), Robot to perform underwater surgery, *National Post* (Canada), 7 April, 2006.

Bouten, C. V. C., Koekkoek, K. T. M., Verduin, M., Kodde, R. and Janssen, J. D. (1997), A triaxial accelerometer and portable data processing unit for the assessment of daily physical activity, *IEEE Transactions on Biomedical Engineering*, 44(3):136–147.

Brenner, R. and Bartholomew, L. (2005), Communication errors in radiology: a liability cost analysis, *Journal of the American College of Radiology*, 2(5):428–431.

Etter, P. C. (2003), *Underwater Acoustic Modelling and Simulation: Principles, Techniques and Applications*, 3/e, Taylor & Francis, ISBN 0419262202.

Felt-Lisk, S. (2006), New hospital information technology: is it helping to improve quality?, *Mathematica Policy Research* 3:1–4.

Fong, B., Fong, A. C. M., and Hong, G. Y. (2005a), On the performance of telemedicine system using 17 GHz orthogonally polarized microwave links under the influence of heavy rainfall, *IEEE Transactions on Information Technology in Biomedicine*, 9(3):424–429.

Fong, B., Fong, A. C. M., Hong, G. Y., and Ryu, H. (2005b), Measurement of attenuation and phase on 26 GHz wideband point-to-multipoint signals under the influence of tropical rainfall, *IEEE Antennas and Wireless Propagation Letters*, 4(1):20–21.

Hirata, A., Fujiwara, O., Nagaoka, T. and Watanabe, S. (2010), Estimation of whole-body average SAR in human models due to plane-wave exposure at resonance frequency, *IEEE Transactions on Electromagnetic Compatibility*, 52(1):41–48.

Li, H. B. and Kohno, R. (2008), Advances in mobile and wireless communications, *Lecture Notes in Electrical Engineering*, 16(4):223–238, Springer Berlin Heidelberg.

Mchugh, T. (1997), MedLink bails out in-flight emergencies, *Phoenix Business Journal*, 21 November, 1997. http://phoenix.bizjournals.com/phoenix/stories/1997/11/24/focus5.html

Martinez, A. W., Philips, S. T., Carrilho, E., Thomas, S. W., Sindi, H., and Whitesides, G. M. (2008), Simple telemedicine for developing regions: camera phones and paper-based microfluidic devices for real-time, off-site diagnosis, *Analytical Chemistry*, 80(10):3699–3707.

Means, D. L. and Chan, K. W. (2001), Evaluating compliance with FCC guidelines for human exposure to radio frequency electromagnetic fields, additional information for evaluating compliance of mobile and portable devices with FCC limits for human exposure to radiofrequency emissions, *FCC Supplement C (Edition 01–01) OET Bulletin* 65 (Edition 97-01):1–53.

Morris, S.J. and Paradiso, J.A. (2002), Shoe-integrated sensor system for wireless gait analysis and real-time feedback,

24th Annual Conference and the Annual Fall Meeting of the Biomedical Engineering Society Conf. Proc., Vol. 3 pp. 2468 – 2469, 23–26 Oct. 2002.

Mullaney, T. J. and Weintraub, A. (2005), The digital hospital, *BusinessWeek* USA, 28 March, 2005.

Pappas, I. P. I., Keller, T., Mangold, S., Popovic, M. R., Dietz, V., Morari, M. (2004), A reliable gyroscope-based gait-phase detection sensor embedded in a shoe insole. *IEEE Sensors Journal*, 4(2):268–274.

Park, S. and Jayaraman, S. (2003), Enhancing the quality of life through wearable technology. *IEEE Engineering in Medicine and Biology Magazine*, 22(3):41–48.

Randerson, J. (2008), Eye-controlled robot may make heart surgery safer, *The Guardian*, 22 March 2008.

Tachakra, S., Banitsas, K. A. and Tachakra, F. (2006), Performance of a wireless telemedicine system in a hospital accident and emergency department, *Journal of Telemedicine and Telecare*, 12(6):298–302.

TriData Corporation (2005), Current Status, Knowledge Gaps, and Research Needs Pertaining to Firefighter Radio Communication Systems, a report prepared for the NIOSH http://www.cdc.gov/niosh/fire/pdfs/FFRCS.pdf

Wang, Q., Tayamachi, T, Kimura, I. and Wang, J. (2009), An on-body channel model for UWB body area communications for various postures, *IEEE Transactions on Antennas and Propagation*, 57(4):991–998.

Welch, T. B., Musselman, R. L., Emessiene, B. A., Gift, P. D., Choudhury, D. K., Cassadine, D. N. and Yano, S. M. (2002), The effects of the human body on UWB signal propagation in an indoor environment, *IEEE Journal on Selected Areas in Communications*, 20(9):1778–1782.

Winters, J. M. and Wang, Y. (2003), Wearable sensors and telerehabilitation: integrating intelligent telerehabilitation assistants with a model for optimizing home therapy, *IEEE Engineering in Medicine and Biology Magazine* 22(3):56–65.

4

Technologies in Medical Information Processing

We so far have looked at a number of situations where telemedicine and related technologies can save lives where a few decades ago this would have been simply impossible. Telemedicine covers just about all corners of the globe. Its comprehensive range of service facilitates everything from search and rescue operations to general health monitoring. All these involve medical information being captured and converted into the digital domain. Numerous advantages exist for handling digital data instead of leaving everything in the original analog form, as (Haykin, 2006) describes: the ease of transmission, processing and subsequent storage with digital data compared to analogue data manipulation.

So, what do these long strings of '0's and '1's representing medical data have that differ from anything else digital in daily life, like CDs and cameras? One thing in common is that in all these applications, information is sent and processed in *binary bits*, i.e. we only deal with '1's and '0's. However, the requirements for capturing and handling medical data are quite different from those for general purpose consumer electronics devices. For a start, medical information is often specifically related to a single individual. A person's medical history must be kept in strict confidence at all times. Compare the consequence of losing a few songs on an MP3 player and losing the analysis results following a medical test. The maximum penalty of the former is probably buying a new CD (if it is no longer in the person's possession) whereas the latter can lead to lengthy legal proceedings and damage claims, and the patient may lose precious time on receiving prompt treatment, in addition to the impact on the medical institution's reputation. The fundamental difference in requirements extends to the way information is processed and tolerance to faults, errors and omissions. Looking at the above comparison again, data misinterpretation may lead to momentary disruption to music playback or degradation in sound quality and there will be no consequences once normal playback is resumed in a few seconds' time. What could happen to the the loss or corruption of medical data can open up a whole nightmare of disasters, including possible failure to diagnose life-threatening conditions.

The course of making use of medical information, just like most information systems as illustrated in Figure 4.1, commences by data acquisition from various sources such as diagnosis

Telemedicine Technologies: Information Technologies in Medicine and Telehealth Bernard Fong, A.C.M. Fong, and C.K. Li
© 2011 John Wiley & Sons, Ltd

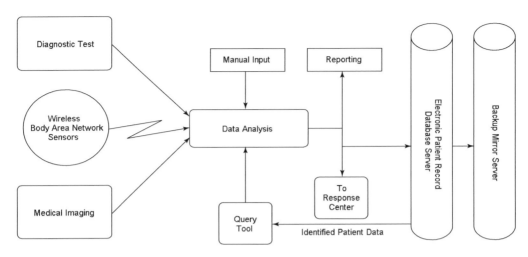

Figure 4.1 Block diagram of a medical information system

and continual monitoring. In the case of telemedicine, the majority of data comes from patients and involves a diverse range of data types from biosignals to surveys about daily activities that require manual entry. Once captured, the data needs to be transmitted to an appropriate location for processing in order to make sense of what the data conveys about the patient. Next, processing entails technologies in different areas such as signal processing, multimedia and data mining; how the data is processed depends on the nature of data and related application. Having analyzed the data such that any necessary actions can be taken in response to the given situation, the data needs to be stored for archival as it can be very useful in a number of ways; for example, a patient who is allergic to certain substances needs to make oneself known prior to receiving treatment. Data can also be used anonymously for statistical analysis of virus mutation and spread pattern in the study of disease control, government agencies can use the anonymous data for regulatory planning, etc. So, an effective way of storing a massive amount of data and speedy retrieval of relevant data is also an important topic to study. The main purpose of this chapter is to walk through the entire process of medical information processing and we shall conclude the chapter by taking a look at the Electronic Drug Store which utilizes medical information for the efficient and safe dispensing of medication. It is an example demonstrating the importance of technological advances in medical information technology for assisting patients with special needs so that medication becomes risk-free and easy to access.

4.1   Collecting Data from Patients

There are all kinds of data to be collected from a patient, from head to toe, within and around the body. We shall concentrate our discussion on biomedical data related to the human body and leave the survey (verbal and written) collection topic behind in order to deliberate on the technical aspects. So, we shall look at what kind of information about a patient can be collected and see how it can be collected. We will also look at an overview of any necessary precautions in the process of collecting such data. The human body is so complex that it would be impossible to cover every single measurable parameter in a single book volume. Our main objective here is to look at some commonly used attributes and to get a good understanding about what is involved whilst processing medical information.

The obvious candidates are vital signs of a human body as these are signs that determine the health state of an individual. Indeed, a person without all of these may not even be alive. We shall look at some of the properties about these signs and how they can be collected. Some of these signs are inherently known to present circadian rhythms in a 24-hour behavioural cycle with fluctuation due to temporal regulation of the ambient environment and activities.

4.1.1   Body Temperature
(Normal Range: 36.1–37.5 °C)

The 'normal' body temperature of a person varies not only based on the surrounding environment, but to a greater extent on where within the body the temperature measurement is taken. (Mackowiak, 1992) reveals that even gender plays a role in the mean body temperature that is considered normal. Body temperature measurement is the principal factor that indicates whether a person suffers from hyperthermia or hypothermia upon exposure to extreme conditions. The former is above 40 °C that may result in severe dehydration caused by excessive sweating; whereas the latter is below 35 °C after exposure to 'freezing' conditions in the cold. Both can be fatal if medical attention is not given promptly. Abnormal body temperature can also indicate fever, which may lead to permanent organ impairment or even mortality. Precise measurement of body temperature and monitoring its changing pattern is therefore an important issue to consider.

There are many methods of measuring body temperature with varying precision and time required for measurement. Measurement can be taken from many points of the body, most commonly armpit, mouth underneath the tongue, ear, or rectum. These positions are listed in ascending order of nominal temperature that spans across the 37.6–38.0 °C range. A number of factors that affect the reading taken are outlined in the study by (Sandsund, 2004). The age of the subject also makes temperature measurement less predictable, children playing hard may generate a considerable amount of heat inside the body as an absolutely normal response whilst the elderly may not have adequate energy to generate as much heat under normal situations. To illustrate the extent of normal body temperature variation during the day, we take a look at a sample reading from three perfectly healthy persons at ages 5, 35, and 70 in Figure 4.2. Although the activities undertaken by each subject varies during the day, the circadian rhythmicities of everyone involved appear fairly consistent. The significance of this behaviour tells us that body temperature measurement for what is considered 'normal' can be quite capricious.

Body temperature measurement can be accomplished in several ways and each method can be affected by different environmental variables. For example, the very traditional way of oral measurement by putting a thermometer into the mouth can result in very significant deviation following the consumption of either hot or cold drink. Likewise, a reading taken under the arm can be greatly affected by sweat and ambient temperature change. Therefore, more reliable methods are developed over advances of technology. For example, tympanic temperature can be measured economically and reliably with an infrared ear thermometer that

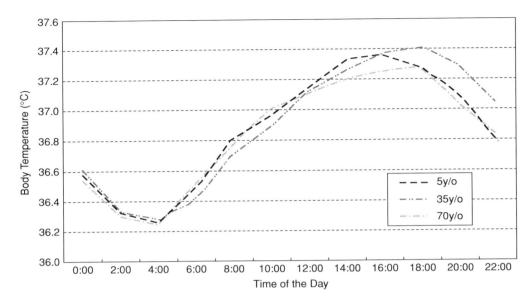

Figure 4.2   Normal variation of body temperature throughout the day

operates by measuring the amount of infrared energy radiated from the subject's eardrum. Ear measurement is intrinsically reliable since the eardrum is situated very near the *hypothalamus*, the core temperature regulator of the human body. This method is fairly fast with a reading obtained in about 0.1s and the small portable thermometer is very affordable for consumer use. With appropriate wireless technology, the reading can be automatically transmitted to a nearby workstation for patient record updating.

The infrared ear thermometer is good for measuring individual person's body temperature as a probe needs to be placed inside the ear for each measurement. To monitor and prevent the spread of certain diseases, body temperature monitoring is sometimes imposed at checkpoints where people come and go. For example, during the SARS and avian influenza pandemics, we saw border control authorities imposing body temperature checks in many countries. To ensure the smooth flow of people traffic, a colour image of each subject is captured by a contactless heat sensing camera that instantaneously shows the core body temperature as soon as the subject walks past the camera. Infrared thermal imaging is commonly used for this purpose where abnormalities of body temperature can be revealed by a change of colour in the image. Although representation of colour does not offer high measurement precision, it is a fast and convenient method that can be programmed to trigger an alarm if a certain colour that represents a certain preset threshold temperature is detected among a group of people who enter the camera's operating area. However, its reliable use demands precise calibration, both in terms of the process of performing instrumental calibration and the calibration stability that determines how frequently the device needs to be calibrated again. Also, its reliability can be significantly affected by surrounding error sources such as radiation and heat generating machinery.

Forehead and spot infrared thermometers are also commercially available although not widely used for good reasons. Forehead measurement can be greatly affected by ambient temperature as well as the use of fever-lowering medication such as acetaminophen or ibuprofen; whereas spot infrared involves the use of a laser beam that can be potentially hazardous if accidentally pointed at a subject's eye.

Accurate detection of high body temperature in infants is a particularly important issue as permanent disabilities can result if treatment is not provided at once. (Cranston, 1975) describes the human body response to infection that results in a fever. The cause of temperature elevation can be simply wearing too much as many parents tend to over-protect little babies. Sometimes this can be a normal response to a vaccination, or, in more serious cases, caused by viral infection that requires immediate medical attention. Technology can help parents monitor their newborn child with a small heat sensing camera in the event of suspecting a fever, the system generates an audible alarm should the baby's body temperature exceeds 38.0 °C and will automatically alert the clinic should the temperature reach 38.9 °C which indicates that the baby requires immediate medical attention. This is an example of where a simple thermometer can be linked to a telemedicine system for improved healthcare monitoring.

Technological advances provide more precise means of measuring body temperature than traditional mercury thermometers, with added feature enhancement such as automatic update of patient records and alerts for temperature exceeding a certain preset threshold. Analysis of temperature variation pattern can also suggest a possible cause that requires medical attention; different measurement methods compromise in terms of speed, precision, and ease of operation. Different methods are optimized for specific applications and operating environments.

4.1.2   Heart Rate
(Normal range at rest: 60–100 bpm)

Measurement and subsequent analysis of heart rate is useful in many applications, from life-threatening conditions such as abnormal behaviour due to heart failure, to general fitness assessment in gymnasiums as discussed earlier in section 3.5. Not as homogeneous as body temperature, the heart beat daily pattern of a human body also exhibits a certain degree of circadian rhythm as shown in Figure 4.3. It shows that generally neglecting any irregular heavy activities, the heart beats at almost 30% higher during the day than sleeping at night and a daily average of around 70 beats per minute (bpm) over a range of around 58–82 bpm. Note, incidentally, that under normal circumstances, a female subject generally beats some 5% faster than male in identical situations. Obviously, more blood is pumped across the body while a person moves around than sleeping. A reading taken once every two hours would eliminate any sudden impulse caused by exercise during the day or a nightmare during the night. As the purpose of obtaining a set of heart rate readings is usually associated with study of certain activities, it would have a much better margin of error than body temperature measurement which is more prone to uncontrollable environmental conditions.

As the measurement unit 'bpm' suggests, heart rate is measured by the number of beats within any one minute period. The easiest way is to count the number of pulses over a one minute interval. In theory, this can be read from anywhere in the body with an artery running near the skin. Most commonly, measurements are taken at the radial and carotid artery (wrist and neck, respectively). Some fitness equipment may use the brachial artery (elbow) for ease

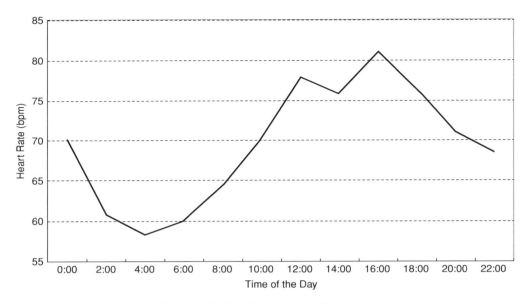

Figure 4.3 Circadian rhythm of heart beat

of access during a workout. Gymnasium equipment is often fitted with heart rate sensors such as the one illustrated in Figure 4.4. As we use the same kind of measuring method on a regular basis, it appears to be extremely effortless for us to use. However, its design is problematic, in that we need to grab the sensor fairly tightly in order to get an instantaneous reading of the heart rate while exercising. We may overlook two important limitations: restriction of movement while gripping the handle and running (as in the case of a treadmill), and lack of means for keeping a log on our health conditions (an instantaneous reading may not convey a lot of useful information about the health state). To improve on the capability, a simple wearable counter can be deployed as follows: an electrical signal is induced through the heart muscle during contraction as the heart beats. A wearable transmitter that picks up the signal can be placed near any of the three locations mentioned above. The transmitter then sends an electromagnetic signal corresponding to each pulse to a receiver that counts over a certain period of time, say five seconds, and displays the heart rate by normalizing it to the number of beats per minute; in the case of counting in five seconds this would mean multiplying the counts by twelve for an estimate of the number of pulses per minute. One of telemedicine's many features is keeping track of health. Based on (Londeree, 1982) that suggests the maximum heart rate for a given age decreases by 1 bpm for each increase of age by one year. A user can program the fitness monitoring device to alert the user to slow down when the heart rate reaches a certain predetermined level to ensure a safe workout. Devices designed for use by the elderly should avoid measuring at the neck as inappropriate exertion of force during the process may risk light-headedness that may lead to serious consequences. For infants of less than one year old, normal heart rates are notably higher than toddlers, hence the range of measurement will be very different. So, different requirements are present for application in different areas.

Figure 4.4   Heart rate sensors

Ultimately, telemedicine technology should assist with getting necessary attention when a sudden change of heart rate may indicate a serious medical situation. There are numerous possible causes of heart beat aberration; hypothyroidism and influence by medication are common causes of lowering the heart rate. Conversely, factors such as heavy exercise, stress, disease, and stimulants such as coffee and alcohol can rapidly increase the heart rate. A simple automated system such as that illustrated in Figure 4.5 can help ensure that elderly people who live alone are monitored at all times. This system is easy to set up and it does not require any user interaction. Quite simply, a small pulse counter is placed at the back of a wristwatch which continuously monitors the user's heart rate when worn. Once the reading falls outside the predetermined nominal range it will send a signal to the responding unit that in turn alerts the service centre via a telemedicine network. A team of support personnel will attempt to place a phone call to find out if the user is undertaking normal activities and immediate attention will be sent to the user in case the call is not answered. One important consideration with such a system is that the threshold must be set with sufficient margin to minimize the chance of a false alarm while any serious problems will not go undetected. However, weak pulse cannot be detected simply by counting the number of beats. Abnormally weak pulse may be due to potentially fatal causes such as blood clot or heart and peripheral arterial disease. It is therefore necessary to deploy more refined methods for measuring heart beats for detecting palpitations.

ECG/EEG (see section 4.2) can also be used for measuring heart rate, these provide accurate measurements and also provide an indication about the rhythm corresponding to heart beat pattern as well as strength. Such additional information can be useful when detecting any signs of heart disease and abnormalities of blood vessels. More sophisticated methods of analyzing heart beats are therefore necessary because counting the number of beats per minute may not be able to provide sufficient information to determine if blockage of blood vessels

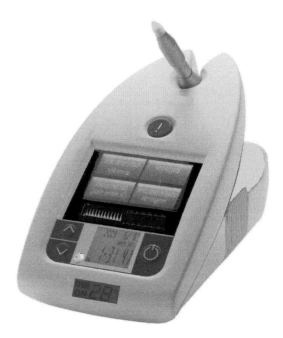

Figure 4.5   Elderly assistive device with environment sensing and communication capabilities

has occurred. Palpitations can be chronic or acute with varying consequences and each have different requirements for detection. (Malik, 1996) describes a number of alternatives such as statistical, time and frequency domain methods. These rely on a series of ECG recordings that may span a prolonged period of time, such as over a 24-hour duration. The main purpose is to distinguish problems from beat irregularities due to normal variations.

Sometimes, a patient's medical history can reveal impending problem areas. Also, there are situations where chemical analysis may be involved when identifying ingestions related to causes of palpitation. For example, detecting the presence of substances that may influence heart beat will normally require laboratory diagnosis before any appropriate adjustments to the measured data can be made. Both situations require linking the testing site to respective departments of the hospital in order to facilitate outpatient palpitation monitoring. It may even be necessary to utilize implanted devices underneath the patient's skin when permanent monitoring for certain conditions is required.

4.1.3   Blood Pressure
(Normal systolic pressure range: 100–140 mmHg)

Blood pressure is a measure of the pressure (force divided by the surface area) exerted on the walls of arteries. It makes blood circulate through the body so that oxygen and nutrients can be delivered to all organs. Compared to the two vital signs discussed above, blood pressure is one that exhibits the least regular pattern over the day. An attempt to plot the blood pressure variation of a certain person during a typical day would most likely turn out to be a messy chart that has little significance. In general, it would be true to say that the blood pressure is normally higher when awake than asleep. This is, however, not necessarily always the case. For a normal healthy adult, the mean systolic pressure (peak pressure in the arteries) is around 120 mmHg but can vary about 20 mmHg above or below the mean during normal activities. Note, incidentally, that the diastolic blood pressure (minimum pressure in the arteries) of the same person is typically slightly over half that of the systolic value with a healthy range of around 80–90 mmHg. Due to the irregular pattern exhibited throughout the day, a spot measurement of an instantaneous reading may not be useful at all. We may recall sometimes when we visit the doctor and they use a *sphygmomanometer*, a very simple non-invasive measurement method where one arm is bound tightly in a cuff and the doctor uses a mechanical hand bulb pump to obtain a spot reading of our blood pressure at the time of the visit.

This seemingly simple task of measuring blood pressure with a traditional method may not have any obvious link to technology. Before we look at how technology sets in let's take a closer look at what blood pressure measurement is about. Quite simply, it is a measurement of the pressure exerted inside a blood vessel when the heart is beating and pumping blood through the arteries of the human body. Such measurement is known as the systolic pressure. This can essentially be done wherever there is an artery near the skin. The diastolic pressure also needs to be measured in most circumstances and is the pressure when the heart is at rest in between two consecutive beats. A hypertension condition is defined if any one of these parameters is too high. To measure the two blood pressure parameters using the traditional method with a sphygmomanometer, usually used in conjunction with a stethoscope for listening to the heart beat so that readings can be taken at the appropriate times; either a pulse is heard for the systolic pressure or the absence of a pulse corresponds to the diastolic pressure. In this manual process, the reading is taken at the moment when synchronized with hearing hence there will certainly be a delay that introduces some kind of error. This is just a spot reading taken at a certain time during a visit to the clinic and is therefore inappropriate for ambulatory blood pressure monitoring (ABPM) that involves continuing measurement throughout the day. This would require a wearable monitor that collects blood pressure readings throughout the day and is able to transfer the data to an external device for analysis by medical personnel. ABPM is usually deployed on a temporary basis for circumstances such as abnormally high blood pressure under the influence of certain prescribed drugs, or patients subject to prolonged anxiety undergoing psychological treatment.

So, there are circumstances when continuous monitoring becomes necessary. This is where technology makes it possible and the subject feels comfortable with the wearable device, particularly useful for hypertension patients who are resistant to pharmacotherapy. The entire process involves reading, scanning, and analysis of captured data. (Marchiando, 2003) has described a number of methods for carrying out ABPM where the appropriate measuring apparatus can be remotely linked to the hospital for off-site measurement. This is particularly useful in monitoring cardiovascular patients where accurate measurement that reflects readings from normal daily activities is necessary. As (Pickering, 1999) explained, many patients tend to become too nervous which drives up the blood pressure reading unintentionally during a doctor visit. Remote measurement would ease tension hence more accurate measurements can be obtained.

Small wearable automatic blood pressure meters are readily available in the consumer electronics market in different forms for measurement taken at different locations of the body.

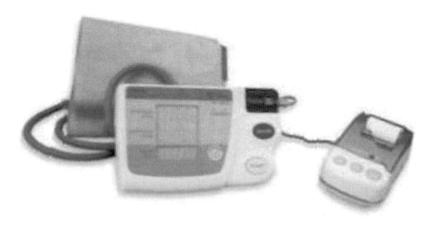

Figure 4.6 Blood pressure meter

In addition to the arm, measurements can also be taken at the wrist, leg, or even finger. An example of a small monitor is shown in Figure 4.6. Many similar devices are on the market priced below 65 Sterling Pounds (US$100). Its design is very simple, air is pumped into an inflatable wrap with a pressure sensing switch that acts like the manually inflated cuff of a sphygmomanometer. A quick succession of readings on the pressure exerted on the switch is taken with one high and low reading that correspond to the systolic and diastolic pressure, respectively. With an internal clock, the time of the reading can be recorded and the data can be stored and printed for analysis.

Telemedicine technology can do more than facilitating remote and periodic blood pressure monitoring. It can also help alert medical personnel when certain methods are not suitable to be carried out on patients with special conditions. For example, applying non-invasive measurement with a sphygmomanometer to sufferers of sickle cell anemia is not recommended since excessive pressure applied to the patient's arm can lead to intravascular sickling resulting in intravascular thrombi, tissue necrosis and haemolysis. To prevent such problems from arising, retrieval of medical history from the electronic patient record can alert medical personnel to the existing condition prior to putting the patient under unnecessary risk by using a sphygmomanometer.

We have looked at a brief description on how technology can assist with blood pressure measurement and monitoring and will now move to the next vital sign, respiratory rate.

4.1.4   Respiration rate
(Normal range: 12–24 breathes per minute)

Among all body vital signs, respiratory rate is probably the most difficult to measure due to its significant variation over a very short period of time. Its pattern is somewhat related to the change in heart rate as the intensity of activity would affect both parameters. Taking a deep breath may lengthen the duration of a breathing cycle thereby reducing the respiratory rate while heart beat is much less affected. Respiratory rate is much lower than heart rate, typically

Oxygen Pressure Sensor
(Internal)
Voice Communication

Respiration Rate Counter

Emergency Button

Beacon

Water Pressure Sensor
(External)
Heart Rate Monitor

Oxygen Saturation Sensor

Figure 4.7   Telemedicine under water

a healthy adult breathes around 12–24 times per minute. The rate varies quite considerably over age: newborns may have over 40 breaths per minute as normal behaviour and the average rate for a toddler may be reduced to around 30. Although respiratory rate may provide less important information than those three listed above when determining the health state of a person, accurate measurement of respiratory rate would be most useful for activities such as diving where the respiratory rate would govern how long a diver can be submerged for. A range of equipment can be fitted to a diver as shown in Figure 4.7, where the most important device is a button for seeking help. Used in conjunction with a beacon, the diver's position can be easily located. As this sub-section is all about technology for measuring respiratory rate, we shall concentrate our discussion on the part which measures the respiratory rate to provide a constantly updated estimate of how much oxygen is left before the diver must decompress and return to the surface. It also triggers a remote alarm to alert support staff on the shore and nearby divers in case sudden abnormal respiration is detected.

Under normal circumstances, respiratory rate is measured for patients with lung disease or taking medication that suppresses respiration. Also, asthma symptoms are closely linked to bouts of breathlessness which can be readily detected by respiratory rate monitoring. *Tachypnea*, the anomalous increase of respiratory rate, is an important behaviour to detect since it can be caused by serious problems such as pneumonia, fever, and congestive heart failure. Breathing is easy to count as it is usually slow and rhythmic, counting the number of expansion and contraction of the thorax can measure the respiratory rate. So, the thoracic motion during breathing can be measured by placing a pressure-sensitive switch with a counter inside a vest. The chest expands as the diaphragm muscle contracts, and the chest cavity shrinks as the diaphragm muscle settles. The frequency of this repeated motion can be counted via the switch.

4.1.5   Blood Oxygen Saturation
(Normal range: SaO_2: 95–100%, PaO_2: 90–95 mmHg)

Blood oxygen saturation measures the ability of the lungs to supply oxygen to the blood. In the blood, oxygen is carried chemically in haemoglobin and dissolved physically in plasma.

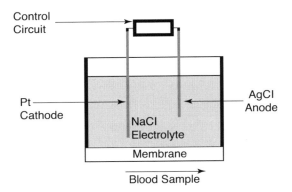

Figure 4.8   Partial pressure of oxygen in arterial blood (PaO₂) measurement

Measurement is done to evaluate the oxygenation and saturation of haemoglobin in the blood. There are several parameters involved, partial pressure (in mmHg) of oxygen in arterial blood (PaO_2), which is a method used to measure the arterial percentage of blood; whereas SaO_2 and SpO_2 refer to direct and indirect measurement of the percentage of the blood oxygen saturation level, respectively. The former is measured by pulse oximetry and the latter is measured by arterial blood gas sampling. Although SaO_2 and SpO_2 may sound similar, these two parameters differ fundamentally. Conditions such as thrombolysis and influence by anticoagulant medications can significantly affect the readings obtained in an arterial blood gas sampling. These parameters are related to respiration as *inhalation* brings oxygen into the lungs while *exhalation* brings carbon dioxide out.

PaO₂ is about gas measurement that can be measured by polarographic oxygen electrode as illustrated in Figure 4.8. It consists of a platinum cathode and a silver chloride anode where an electrical current is generated which is proportional to the oxygen tension. The blood sample is isolated from the electrode by a membrane to avoid protein deposition. The apparatus has to be kept in a temperature-controlled oven in order to maintain a temperature similar to that of the human body of around 37 °C. Another precaution is to ensure that the membrane does not have any protein deposit that may accumulate on its surface over time.

Pulse oximetry is a non-invasive method of continual arterial oxygen saturation monitoring. Pulse oximeters are usually small portable devices that paramedics can carry to attend an accident scene. These can measure the arterial oxygen saturation (SaO_2) of a patient. In theory, the maximum amount of oxygen that the blood can carry can be calculated from Equation 4.1 below:

$$Sa\,O_2 = \frac{O_2 content}{O_2 capacity} x\,100\% \tag{4.1}$$

This would give some insights into what to expect on an oxyhemoglobin dissociation curve. More accurate measurement of the actual value needs an oximeter that relies on a light source with red and infrared LEDs (600 and 800 nm wavelength, respectively) that gleams through certain parts of the body where a relatively translucent area of blood flow can be exposed to the light. Oxygenated haemoglobin absorbs infrared light whereas deoxygenated haemoglobin

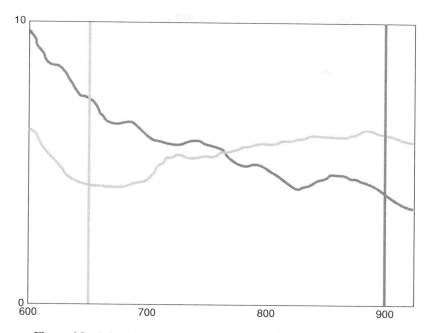

Figure 4.9   Infrared energy absorption by hemoglobin versus wavelength

absorbs red light as illustrated in Figure 4.9. Measurement is very often taken from the finger or ear lobe. Light passes through the blood vessel that absorbs a certain portion of red and infrared light beam. Whatever is left over is received by a photocell that can then deduce the red-to-infrared ratio of absorbed light through blood. This simple arrangement is shown in Figure 4.10 where a 100% SpO_2 yields a received light ratio of about 0.5. It should be noted that calibration is necessary due to the varying extent of light absorption by skin and tissue.

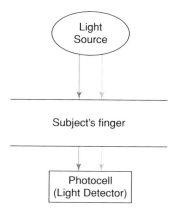

Figure 4.10   Pulse oximeter oxygen saturation (SpO_2) measurement

Also, the amount of arterial blood flow varies due to heart beat sequence that may affect the measurement accuracy. It is therefore necessary to measure for a sufficient time covering two successive heart beats to obtain an average reading. The measurement of oxygen saturation at an accident scene is important for detecting hypoxia so that necessary emergency treatment can be provided by the time the patient reaches the hospital. It is also worth noting that conditions such as tricuspid regurgitation, hypovolaemia or vasoconstriction affecting blood flow may impair the reading from an oximeter. As a final note, an oximeter cannot distinguish carboxyhaemoglobin from normal oxygen-carrying haemoglobin in the event of carbon monoxide poisoning and the reading obtained may be higher than what it should actually be.

4.2   Bio-signal Transmission and Processing

The main function of telemedicine is to provide medical services remotely. To serve this purpose, data must be transmitted from one location to another, such as from an accident scene or a patient's home to the hospital. Further, any data received needs to be processed before any useful information can be extracted for analysis and storage. There are so many types of relevant information. Some are fairly self-explanatory like instructions for taking medication, whereas parameters like oxygen saturation may require expert analysis before the cause of any abnormalities can be established.

For any kind of data about a patient to be collected and processed, we need some kind of mechanism similar to that of Figure 4.11, which is expanded from the basic communication system shown in Figure 2.1, with inherent additive noise in Figure 2.2 understood and omitted for simplicity. Here, we have a simple block diagram showing biosensors that capture data, such as those described in section 4.1; the sensor network, being connected to a transmitter, via an analog-to-digital (A/D) converter, will send the collected data to a remote receiver. The purpose of converting the captured analog data into digital domain is for transmission efficiency and security. While transmission efficiency will be dealt with in this section, the topic of information security will be addressed in Chapter 6. At the receiving end, the data will be analyzed and/or stored. Stored data can also be retrieved for analysis at any time.

This is a typical information system that deals with basic information theory. It would be virtually impossible to discuss the topic any further without revisiting the landmark work by (Shannon, 1948), which quantifies information in *entropy*, a term that refers to a certain

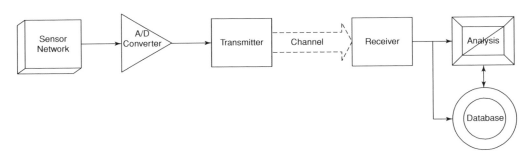

Figure 4.11   Block diagram for collecting patients' information

anticipated value in association with a data set. In essence, Shannon entropy measures the maximum amount of information that can be sent across a given communication channel. The theory essentially describes the capacity of a given channel based on a statistical model of the channel under the influence of additive noise during the transmission process. We shall not go into the mathematics behind it, readers interested in the underlying mathematical theories are advised to study the comprehensive reference by (Cover, 2006) for details. Leaving the mathematics behind, the concept is fairly simple. We begin a brief discussion by referring to the basic communication system shown in Figure. 2.1, whose transmitter consists of a discrete source S (with finite number of possible values per output sample) that produces raw data at a rate of R bits per symbol. The source has entropy:

$$H(S) \leq R \tag{4.2}$$

Shannon's Theorem indicates that S can be coded into an alternative, but equivalent, representation at $H(S)$ bits per symbol. The original representation can be recovered in its original form by the receiver. This is theoretically possible as long as the transmission rate is above $H(S)$. Therefore, $H(S)$ is a measure of the actual information content in the output of S. Next, we also look briefly at Channel Coding by considering the transmission of a stream of information bits $b \in \{0, 1\}$ over a digital communication channel with bit-error probability (the probability of having an error bit per one million bits sent) q and capacity $C = C(q)$. A channel code consists of a block of k information bits and maps these bits into a new block of n such that $n > k$ coded bits, c, hence introducing *redundancy*. The *information content* per coded bit r is:

$$r = \frac{k}{n} \tag{4.3}$$

The coded bit sequence c is transmitted and a decoder at the receiver produces estimates $\hat{b}$ of the original information bits, such that the probability of error is:

$$p_b = \Pr(b \neq \hat{b}) \tag{4.4}$$

So, p_b can be minimized given that $r < C$. We can see from the above discussion that C is a measure of the channel quality, that is, how noisy the channel is. With the basic concept of channel quality understood, we shall proceed to the topic of transmitting and processing medical information.

4.2.1 *Medical Imaging*

Medical imaging technology is very widely used in areas such as x-ray, body scan (whole or certain part), anatomy, remote surgery, and accident recovery. Here, we begin by looking at the simple flow chart depicted in Figure 4.12 which shows the process of medical imaging. In almost any situation, medical images are captured, sent, analyzed and stored. In non-emergency cases, images are scanned and stored for later referral or kept for archival purpose. Whereas

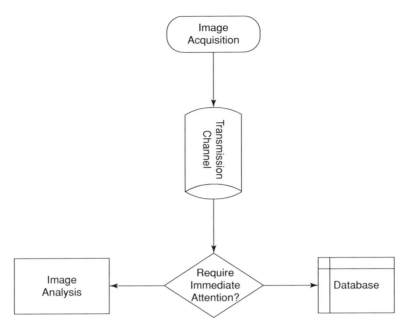

Figure 4.12   Process of medical imaging

immediate attention will be given once the image is obtained in case of an emergency, these are situations such as MRI (magnetic resonance imaging) scans for surgery or photographs taken at an accident scene showing the wounds of an injured patient. Before going deeper into the transmission aspects we first look briefly at how various types of images are taken:

4.2.1.1   Magnetic Resonance Imaging

Shown in Figure 4.13, an MRI scanner looks similar to a tunnel about the length of an adult's body when lying flat, surrounded by a large circular magnet with an RF coil and a gradient coil. The magnet generates a strong magnetic field that aligns protons within the hydrogen atoms. All the protons line up in parallel to the magnetic field like tiny magnets. The radio waves knock the protons from their position when the scanner operates by emitting short bursts of radio waves towards the subject. The subject slides into the scanner during the image acquisition process. When emission stops, the protons realign back into their original random orientations. During this realignment process they too emit radio signals. The protons that locate in different tissues of the body realign at different speeds so that the signal emitted from different body tissues diverges hence tissues of different properties can be identified by such variation of signal emission. From the radio signals, a spectrometer inside the scanner can produce an image based on the body. An example of an MRI scan of a healthy human brain is shown in Figure 4.14. Key features are the different shades of grey that represent a different part of the brain's composition.

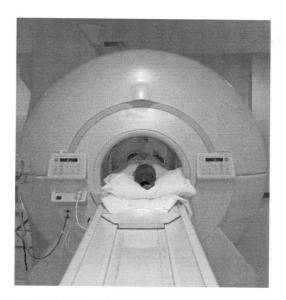

Figure 4.13 Magnetic resonance imaging (MRI) scanner

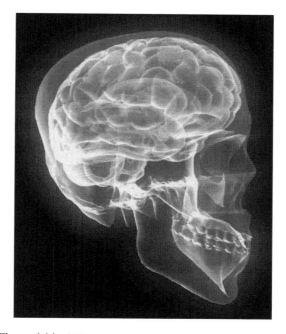

Figure 4.14 MRI scanned image of a healthy human brain

4.2.1.2 X-ray

Similar to an MRI scanner, an X-ray camera is also operated by a radiographer who controls how and where the image is taken. An X-ray image, commonly known as a radiograph, is usually taken for diagnosis purpose. X-ray radiography is perhaps the earliest medical imaging technology that was introduced by Röntgen in 1895 (Koeningsberger, 1988). A portable X-ray camera was made commercially available a year later. This was about a century after Senefelder invented lithography. While X-ray radiography has been very widely used in medical science for over 100 years, lithography never finds its application in medicine. Indeed, the invention of X-ray was such an important event that it won Röntgen the first Nobel Prize for Physics in 1901.

X-ray incurs energy that is sufficient to ionize atoms resulting in positively charged ions that may damage human tissue. X-ray radiography relies on the capturing of radiated electro-magnetic (EM) radiation whose frequency range, hence energy level from elementary physics as in Equation 4.5, way above that of visible light.

$$E = h.f$$
$$h \sim 6.63 \times 10^{-34} \, (J \, s) \tag{4.5}$$

The incident energy E, measured in electron-Volts (eV), is directly proportional to frequency f since h is the Planck's Constant that relates the energy in one quantum. This is the potentially harmful energy that can lead to health problems as such an amount of energy in excess of 1 KeV can change the chemical bonds of vital substances within the human body. Note, incidentally, that radio frequencies do not carry sufficient energy to alter an atom. For this reason, MRI is much safer than X-ray.

The physics behind X-ray radiography is actually quite simple. Consider the situation where an X-ray beam carries sufficient energy to 'knock off' an electron within an atom causing it to ionize, as illustrated in Figure 4.15. An X-ray photon strikes an electron causing the electron to move from a higher energy shell into a lower energy shell closer to the nuclear, this process releases dissipation energy that produces a photon. The photons produced during this process are known as fluorescent or characteristic energy.

To study X-ray image processing, we need to understand how a clear image can be produced. The above physical properties lead to Compton scattering when the incident X-ray photon is deflected from its original path due to an electron. Another Nobel Prize in Physics was awarded in 1927 to Compton for the discovery of this phenomenon. Unlike the above situation, only a part of the photon energy is transferred to the electron during the X-ray strike. So, a photon is emitted with less energy through an altered path. The energy shift caused by reduction of energy, hence wavelength change $\Delta\lambda$ (as in the simple relationship $v = f\lambda$), depends on the angle of scattering:

$$\Delta\lambda = \frac{h}{m_e v} (1 - \cos\theta) \tag{4.6}$$

$$\Delta\lambda = \lambda' - \lambda \tag{4.7}$$

The scattered photon has an energy E' relative to E is:

$$E' = \frac{E}{1 + \frac{E}{m_e v^2}(1 - \cos\theta)} \tag{4.8}$$

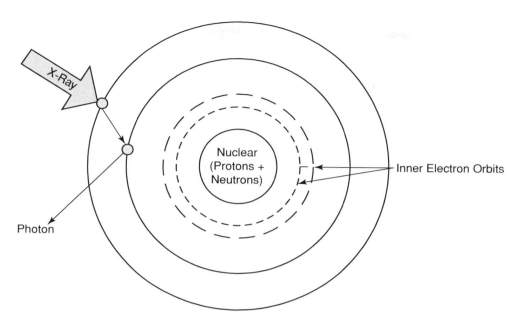

Figure 4.15   X-ray radiography

where m_e is the mass of electron which is a constant and θ is the photon's scattering angle as shown in Figure 4.16, λ' and λ are wavelengths of scattered and incident x-ray photon, respectively. Energy is lost to an electron that is driven out from the atom. Compton scattering is an important topic to study since it is the major source of background noise on an X-ray radiograph. It is also the major cause of tissue damage. It is obvious from Equation 4.7 that the scattered energy E' is independent of scattered angle θ if the incident energy E is low. So, scattered photons with higher energy will continue in about the same direction as that of the X-ray source.

There is a tradeoff between patient's safety and effectiveness of X-ray penetration that produces a clear image in the adjustment of X-ray dosage. The absorbed dose exposed to a patient is measured in terms of the energy absorbed per unit of tissue. Details on X-ray dose can be found from the RSNA (Radiological Society of North America) report of 2009. Other sources of interference include cosmic radiation, nuclear plants, and natural radioactive materials that exist almost everywhere. More details about the possible risk of excessive radiation dosage are discussed in section 8.5.3.

Since X-ray images reveal abnormalities inside the body, small tumours are divulged somewhere inside the image with different shades of grey. Conversion of images into digital format can make transmission and storage far more efficient than with silver-based films. Therefore, preservation of tiny but important details requires digital imaging techniques that provide sufficient resolution and bit-depth that can distinguish any tumour from the background. Additive noise imposed on the image or transmission loss may completely ruin the usefulness of a radiograph. We shall look at the details in sub-section 4.2.2.

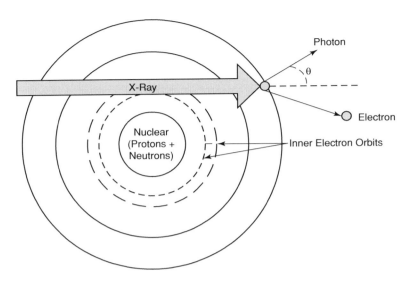

Figure 4.16 Photon scattering

4.2.1.3 Ultrasound

Ultrasound measurement relies on several different properties of sound propagation; these include propagating velocity, attenuation, phase shift, and acoustic impedance mismatch. With variation of these properties while propagating through different substances, tissue structure characteristics can be analyzed (Tempkin, 2009). It is a high frequency sonic signal above the audible frequency range that propagates through fluid and soft tissues. The ultrasonic signal is then reflected back as 'echo' to form an image. The denser the tissue it strikes the more is reflected back producing a lighter image. Images of organs and structures with different shades of grey can therefore be created.

An image is formed by scanning a probe across the area of interest, this probe does not have to enter the body and the entire process is carried out on the skin. The probe emits pulses of ultrasound and picks up the echo as the ultrasound signal is reflected back. We first take a look at how an image is generated by using an example of a heart scan that generates an 'echocardiogram'. The ultrasound signal penetrates through blood in the heart chamber, and is reflected back when it strikes the solid valve. The presence and absence of tissue reflecting the signal produces a black and white image with varying contrast as in Figure 4.17. A monochrome image that shows a healthy heart is formed. This is particularly useful in detecting any abnormalities that may lead to heart problems. Very similar techniques can be used in different areas such as detection of breast tumour and renal calculi (kidney stones) for cancer and hydronephrosis diagnosis at early stages so that early treatment can be provided before the condition deteriorates.

In addition to providing early treatment, ultrasound scan is also very widely used on pregnant women to constantly monitor the development of their unborn babies in the womb. An example of a 21-week-old healthy growing foetus is shown in Figure 4.18. These seemingly blurry pictures convey important information such as gender of the child and whether all parts of the foetus are developing normally.

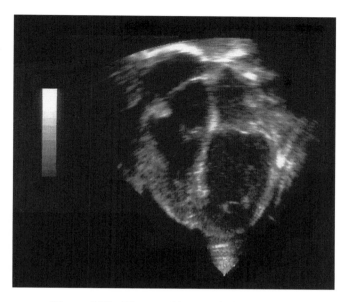

Figure 4.17   Ultrasound image of a beating heart

4.2.2   Medical Image Transmission and Analysis

We have studied three major types of medical image acquisition technology above. We shall now move on to the topic of processing these images without going further into alternatives such as OCT (Optical Coherent Tomography) and PET (Positron Emission Tomography) as

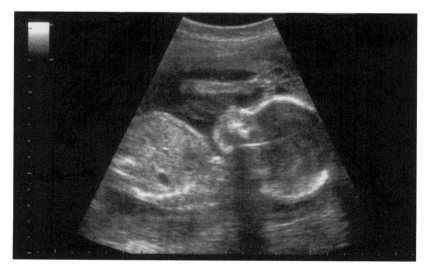

Figure 4.18   Ultrasound image of a healthy foetus

these modalities exhibit many similarities when compared with the image types that we have covered as far as image processing algorithms are concerned.

The technologies related to transmitting a medical image from one location to another may be very similar to that of general purpose photo transmission just like snapping a photo with a camera-equipped 3G mobile phone and uploading a digital photo onto the web. The procedures may be similar but the requirements are certainly very different in the sense that faithful reproduction is the key to an image's usefulness since the main objective of taking a medical image in the first place is likely for identification of any subtle details embedded in the image. It may be a tumour hidden somewhere in a confined area of the image that needs to be identified. Also, many such images are monochrome so the different shades of grey can hold the key to diagnosis from the image.

To learn more about the successful transmission of medical images, we look at a case study where an X-ray radiography is sent from the radiographer's site to an expert for analysis by first referring to (Maintz, 1998). Remember, an X-ray radiograph is a 2-D depiction of a 3-D mapping that represents the attenuation of X-ray absorption properties of tissues that are exposed to a dose of X-ray. In this case, suppose an X-ray beam of intensity I strikes the tissue of a subject; the cross-sectional area of the X-ray beam and that of an atom within the tissue is A and S, respectively. The atom density of the tissue, namely the number of atoms per cubic centimetre of the tissue, is N. The total cross-sectional area of the atoms in this mass of tissue is therefore $N \times S$, and the total area of atoms hit by the beam is $A \times N \times S$. These parameters are shown in Figure 4.19. The rate of change of the beam intensity while penetrating through the tissue across thickness x is:

$$\frac{dI}{dx} = -N.S.I \qquad (4.9)$$

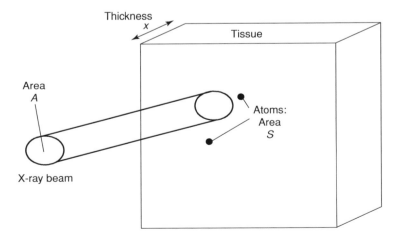

Figure 4.19   When an X-ray beam strikes the tissue

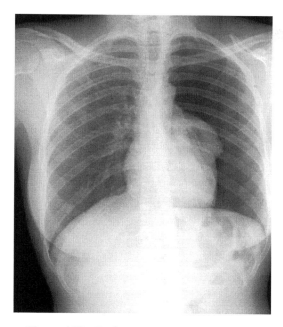

Figure 4.20   Radiograph of tumour in the lung

This is very important in deducing the *attenuation coefficient* μ, which is a function of the photon intensity at any arbitrary position through the tissue x, namely $I(x)$, as:

$$I(x) = I.e^{-\mu x} \tag{4.10}$$

So, the image formed on the radiograph is essentially a map of the photon energy across the area photographed with adequate contrast between bone and different types of tissues. For example a tumour will be highlighted on the radiograph with different a shade of grey compared to bones and healthy tissues. In the sample radiograph of Figure 4.20, the left side of the patient shows an abnormally dark cavity, indicating a decay of tissues inside the left lung. This particular radiograph reveals spontaneous pneumothorax caused by pneumonia; diagnosis is only possible given the clarity of contrast exhibited. This can be compared with the right lung which is perfectly normal and appears much lighter on the image. Successful diagnosis requires an image to be received with relevant details intact; image analysis will become meaningless if the details are lost during any stage of image transmission and processing.

The visual world is composed of analogue images. This sentence makes good sense since images we see in the real world are collections of a continuous spectrum of colours with an infinite amount of details. It is practically impossible to send any image with infinite details. So, the process of digitizing an image would bring it down to a finite size so that sending or storing becomes possible. Transmission of images requires efficient use of available channel bandwidth as a vast amount of data is involved. For example, a simple 'bitmap' (matrix of grey or colour dots called *pixels*) image of 3 000 × 2 000 pixels (i.e. 6 megapixels (MP) resolution) with 256 shades of grey between deep black and pure white, when *uncompressed*, has a file size as:

$$Uncompressed\ bitmap\ file\ size = H.W.2^b \qquad (4.11)$$

Excluding any redundancies such as error checking and additional information about the image including the image type and date taken embedded into the file. Here, b is the number of bits per pixel that gives the levels of shades or colour depth, H and W are the height and width of the image, respectively. In this example, substituting the numbers into Equation 4.11 yields: $3\ 000 \times 2\ 000 \times 8$ ($b = 8$ because $2^8 = 256$ that gives the number of shades) $= 5.72$ MB. The calculation is very simple: we multiply H and W to get the total number of pixels in the image. After multiplying b for the number of bits per pixel we convert the number into units in *bytes* by dividing the product by eight because each byte contains eight binary bits. From there, since each kilobyte (KB) contains 1 024 bytes we then divide the number of bytes by 1 024 to express the size in KB. Similarly, we further divide this number of KB by another 1 024 to express the file size in megabytes (MB) since one megabyte consists of 1 024 KB (but not 1 000 KB). This gives us some ideas about how much data is involved when handling a digital image. It is therefore desirable to make the image smaller for easier transmission and storage.

4.2.3   Image Compression

Compression sets in when shrinking an image for transmission or storage in order to improve transmission efficiency or save space. The problem is that many data compression algorithms are *lossy*; this means some details of the original image are not preserved so that the processed image recovered by decompression is not exactly the same as the original image before compression. Whereas with *lossless* compression, the original image can be converted back into its exact form after decompression without any loss of detail or clarity. This is to say that no difference should be detectable when comparing the two images before and after compression and subsequent decompression. Before going further into this topic, we should remind ourselves that a digital image (one that has been digitalized) consists of an array of pixels that is represented by a long string of '0's and '1's. We begin our discussion by summarizing the pros and cons of lossless and lossy compression methods as reviewed by the tutorial of (Tobin, 2001). Medical image compression is important for improving the efficiency of transmission over telemedicine networks and to reduce the cost of storage for mass electronic patient records.

Colour is represented in digital images by using varying amounts of red, green and blue light; the three primary colours (not to be confused with the definition of primary colours as: yellow, magenta, and cyan; these are classified as 'subtractive colours'). This colour representation of each pixel in a digital image is the same as almost all consumer electronics appliances such as TVs, computers, and cameras. Any colour can be reproduced by adding together percentages of red, green and blue of varying proportion. 'Additive colour' is the process of mixing red, green and blue light to achieve a wide range of colours. In a simple colour bitmap, each pixel is represented by three numbers to store the amounts of red, green and blue light that define the colour of that particular pixel. In such simple bitmap, each pixel requires one byte for each primary colour giving a total of three bytes per pixel. Since one byte contains eight bits, each pixel requires 24 bits to store all the colour information. So, the total number of possible discrete colours this bitmap contains is $2^{24} = 16\,777\,216$, approximately 16 million possible colours. Twenty-four-bit colour images are known as 'True Colour' images in computing terms. The total number of possible colours is given by:

$$Number\ of\ Colours = 2^b \qquad (4.12)$$

where b is the number of bits per pixel or 'bit-depth'. For more faithful and vivid colour reproduction, may consumer digital cameras have 12 to 14 bits per colour.

Compression works by finding areas in an image which are all the same colour, these are then marked as 'this area is all the same colour'. Compression is essentially a process of eliminating gaps, empty fields, and redundancies within an image. The main problem with compressing medical images is that they usually contain a vast amount of subtle and important details, making lossy compression generally not suitable. These details are what restrain areas from being all the same colour or shade of grey, and as such the details can easily be lost due to compression. In many medical images, the details represent very subtle colour and grey-shade variations that may be too subtle to be discernable by a human eye while containing vital information about the health state of a patient. This includes situations such as early development of cancer tumour or foetus with abnormality. Lossy image compression algorithms may involve discarding faint details. The term 'quality factor' is commonly used to describe the extent of image quality degradation. In Figure 4.21, we compare the effect of varying *compression ratio* with reference to an uncompressed MRI scan in (a). Figure 4.21 (b)

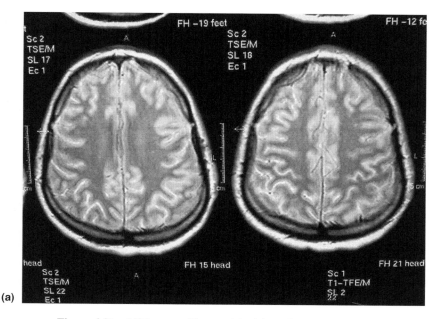

Figure 4.21   MRI scanned image. (a) without data compression

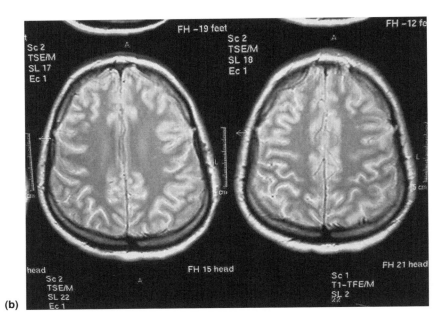

(b)

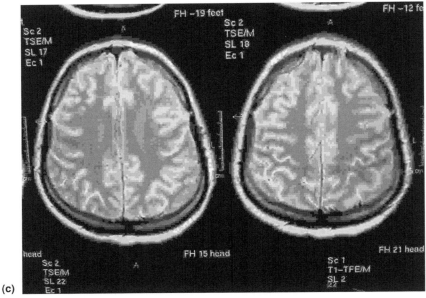

(c)

Figure 4.21   (Continued) (b) moderate compression of 1:20; (c) compressed to 1:100

has undergone a moderate compression of ratio 1:20 and Figure 4.21 (c) has been compressed to 1:100. Is there any noticeable difference? Under close investigation, (b) is a bit more abrasive than (a); and (c) is fairly coarse and blur.

Unlike lossy compression, lossless image compression maps the original information sequence into a string of data bits to reduce the file such that the original image can be recovered exactly from encoded bit stream. Lossless compression does not achieve as high compression ratio as lossy methods so the compressed file size of a same image will be larger.

4.2.4   Biopotential Electrode Sensing

Electrical activities such as Electrocardiogram (ECG), Electroencephalography (EEG), Electromyography (EMG) and graphic hypnograms incur measurement of heart, brain, muscle, and sleep behaviour over time. These are usually measured by the electric potentials on the surface of relevant tissue that correspond to nervous stimuli and muscle contraction over the duration of measurement. These are graphical representations of biomedical waveform generated by plotting electrical current amplitude over time. For the purpose of illustration, we shall concentrate our discussion on ECG data processing as other parameters exhibit very similar properties. Figure 4.22 shows examples of each of these four measurements. One important attribute in common is that all plots are irregular variation of amplitudes over a long measurement time.

ECG records the electrical activity of the heart as it beats. It should be noted that no electricity is sent through the body in the entire measurement process. The electrical impulses made while the heart is beating are plotted so that any abnormal activity with the heart beat rhythm can be identified. A range of possible causes can also be deduced from the plot. ECG is extremely useful in detecting and monitoring problems such as heart attack, coronary artery disease, prevalence of left ventricular hypertrophy and carotid thickening. Numerous sources of noise can impair the measured signal, these include ablation, electric cautery, defibrillation and pacing. Any impulse noise, of excessive amplitude and short duration, can severely affect the detection of abnormalities in the signal. Certain measurement procedures may also affect the effectiveness of ECG measurement. For example, patients suspected of having narrowing of the arteries to the heart may need to undertake ECG measurement while exercising on a treadmill since the plot can appear misleadingly normal if the measurement is performed while the patient remains stationary under such medical condition. Since measurement is taken with electrodes adhered to the chest, movement and shock may affect the accuracy of measurement. Depending on specific application, a measurement session can last as short as one minute, or it can be much longer. A short measurement is likely to be less tolerant to additive noise and interference.

The study of ECG graphs is usually performed manually by a physician. In case the graph is transmitted or stored electronically there is bound to be a loss of quality as in the case of image processing discussed in the previous section. The electrocardiographic patterns need to be reproduced with such clarity to preserve all the useful features. Scanning can be a little difficult because the signal must be clearly separable from the background grid. For this reason, pure black-and-white is not a desirable despite the fact that the plot representing the signal is a monochrome line. Sometimes, separating the image into three separate primary colour channels helps extract the signal from the background grid. The pink grid, which appears only

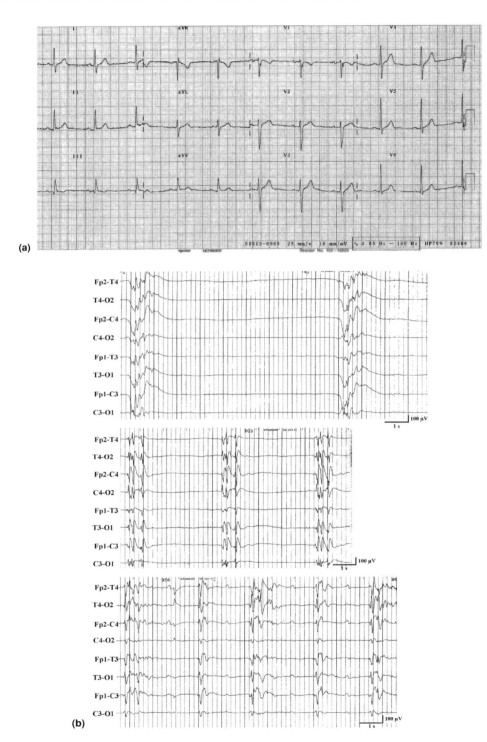

Figure 4.22 Electrical activities. (a) Electrocardiogram (ECG); (b) Electroencephalography (EEG)

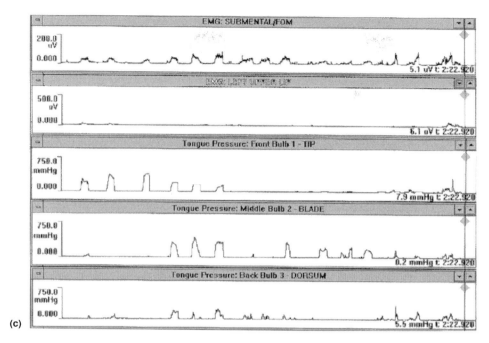

(c)

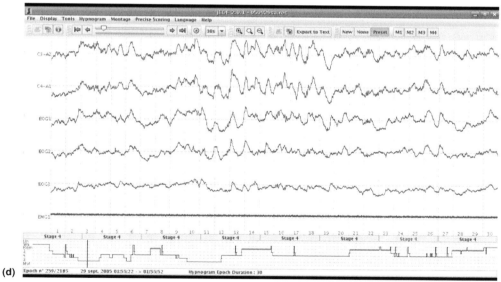

(d)

Figure 4.22 (Continued) (c) Electromyography (EMG); (d) graphic hypnogram

in the red channel, can therefore be easily removed from the plot simply by eliminating the red component of the plot.

4.3   Patient Records and Data Mining Applications

History about a patient's doctor visits have been kept for almost as long as medical science began centuries ago. Legacy paper log cards are still widely seen in many clinics for the sole purpose of recording details of each visit by each patient. There must be a good reason for keeping all these records. First, the patient's conditions over time can be tracked. It can also alert the doctor to conditions such as allergy to certain substances or drugs. Also, repetitive appearance of certain symptoms may indicate something serious. All these can clearly show up on the patient's log card. Private doctors, especially those who have been practising for decades, are so reluctant to switch to electronic patient records since the migration may involve manual data entry of many records. Another deterrent is perhaps the time necessary to get used to a new electronic system, both for updating the records and for information retrieval. They may be so much used to systematically filing paper records by patients' names. A clinical assistant would manually dig out the record of a patient and make it available to the doctor before the visit. The doctor updates the record in writing at the end of the visit and the assistant puts it back into the shelf. This process sounds simple but there are several major problems. First, the doctor or the patient may move, when the doctor ceases practicing due to retirement or whatever reason the records will be left behind. One common question many may ask is whether the writing on those cards is legible. It would be meaningless if the new doctor comes in and is unable to read the information scribbled on the cards. Another major problem is that records just keep on appending. Some patients may have a thick block of log cards, and since it would be difficult to detect who has ceased to be a patient of the clinic some records may just sit on the shelf forever. If the patient has emigrated, the record will be left redundant and there will not be any previous medical history available to the new family doctor.

A rural clinic may have hundreds of patients whereas a large hospital in a metropolis can serve over 100 000 patients. Consider storing medical records for each individual patient from the time of birth, including all test and diagnosis results, prescriptions given, details of each visit. How much data is involved? Each patient may have megabytes of medical history. Those with a long history may even run into gigabytes each. It is not difficult to grasp how vast a medical data bank of a single hospital can be, so how about a national medical database if every single citizen of a country is included? What kind of data backup facility is needed and how can individual entries within the massive database be retrieved quickly and reliably? These are fundamental questions that we need to ask. Although nowhere near the size of the Internet, the information stored is still vast. This is where data mining technology comes in.

Take Mexico City as an example, with a population of over 22 million. An outbreak of swine influenza in March 2009 has driven over 10 000 of its citizens with any flu related symptoms into its hospitals in one single day. These may consist of thousands of unrelated cases, hundreds of suspected cases, and tens of confirmed cases of A(H1N1) infection. These certainly involve a large amount of data if any attempt to keep medical records of all cases is made. Retrieval of informative data for analysis of disease mutation and spread amongst the cluster of data collected on a daily basis is only made possible by data mining technology.

Data mining relies on statistical models for fast retrieval of information from a vast database. One similar application where we frequently make use of data mining is searching over the Internet, for example, using Google™ web search. What we have is a *search engine* that is linked to millions of websites throughout the world. Once we enter the word or phrase to search for, it will grab all pages containing the search string in a fraction of a second. So, how does it work? To facilitate our discussion, we illustrate by searching for a single word for simplicity. Of course, as far as the computer is concerned a phrase is just a very long word with 'space' as a character such that it just treats each space as a letter in a word. Computers understand characters (letters, symbols, spaces alike) in ASCII (American Standard Code for Information Interchange) codes, each character is given a unique seven bit code for identification. For example, an 'A' is known to a computer as '1000001', equivalent to number '65' in decimal representation. So, any word, or indeed phrase, is just a string of ASCII codes or sets of 7-bit codewords entered in a sequential manner.

Data mining involves pattern extraction by examining records in vast *relational databases* from various dimensions and categorizing them. As computational processing power and disk storage capacity increases, more effective statistical analysis software are made possible to search through very large amounts of information within a fraction of a second. To illustrate the power of modern search engines, the authors perform an Internet search for the phrase 'data mining' and results are displayed with over 21 million in a mere 0.18 second. The process analyzes relationships and patterns in records based on open-ended user queries as a search is initiated. Generally, there are four distinctive steps necessary for information retrieval as illustrated in the flow chart of Figure 4.23. Although searching through electronic patient records require very similar technologies, medical records may contain more than simple text and numbers. As we have learned a number of medical image types are also related to individual patients. From real life experience we find that current technology in image search is very futile as very often the majority of displayed results are totally unrelated to what we intend to search for. Essentially, data mining extracts items based on four types of relationships:

- Associations: data are extracted to identify associations or links. For example, patients may be linked between diabetes and obesity as many diagnosed with diabetes are obese. However, it is not necessarily true that someone with diabetes must be obese.
- Classes: data are grouped according to set classes. For example, patients with diabetes can be grouped together in one class.
- Clusters: data are grouped according to logical relationships. For example, patients can be grouped by geographic or demographic criteria. This is particularly useful in studying statistical disease patterns.
- Sequential patterns: data are extracted to predict behavioural patterns and trends. For example, obesity can be linked to chronic disease so that a diabetes patient may be obese, and is more likely to suffer from chronic problems than a patient not classified as diabetes.

We go further by looking at a case study of an electronic patient record for a diabetes sufferer. Vital but sensitive information including name, gender, date of birth, and contact details, is stored. The age alone can give an indication as to classify whether this patient is of diabetes Type 1 or 2, since Type 1 diabetes is usually developed during childhood whereas Type 2 diabetes mainly affects adults of over 40 years old. A large part of electronic patient record

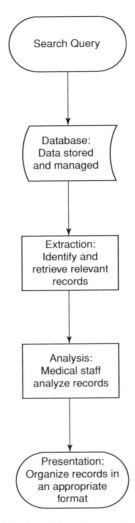

Figure 4.23 The information retrieval process

contains information about each clinical visit such as date of visit, nature of conditions, what remedies have been prescribed since diagnosed and glucose result with respective measurement unit: milligrams per deciliter (mg/dL) or millimoles per litre (mmol/L). Table 4.1 shows a list of countries and unit of measurement. Any follow up and known effectiveness of remedies is also recorded. In addition to text describing the above, digital images and audio recordings may also be included for different situations such as X-ray radiographs and heart beat rhythms for a complete record.

There are methods in searching by pattern recognition as described in (Elmaghraby, 2006), where rule-based techniques are illustrated. In fact, a number of possibilities exist and their

Table 4.1   Blood glucose units

mg/dL	mmol/L
Argentina	Australia
Brazil	Canada
Caribbean Countries	China
Chile	Ireland
Israel	The Netherlands
Japan	New Zealand
Korea	Russia
Mexico	Scandinavia
Most EU States	Singapore
Most Middle East Countries	Slovakia
Peru	South Africa
Taiwan	Switzerland
Thailand	Ukraine
USA	UK
Venezuela	Vietnam

effectiveness depends primarily on the database size and query complexity; as these would demand more computational processing power. The commonly used analysis methods include:

- Neural Networks: predictive computational model that replicates the biological nervous system consisting of many interconnected processing elements as 'neurones'. The model has to be 'trained' through its learning process. So, its performance increases over time given 'adequate training'.
- Data visualization: visual analysis of complex relationships in the data, involving schematic abstraction of data graphically.
- Decision Trees: branch structure that leads to sets of decisions, which derive rules for classification of data records. Two commonly used methods are *Classification and Regression Trees* (CART) and χ-*Square Automatic Interaction Detection* (CHAID); in CHAID, the 'CH' is taken from the Greek alphabet 'χ', equivalent to Chi. These rules are applied to new and unclassified data for data extraction.
- Genetic Algorithms: adaptive heuristic search algorithm that replicates the natural evolution process, it relies on a combination of selection, recombination and mutation to evolve a set of rules. Although everything is based on Charles Darwin's work of the nineteenth century it was first applied to data mining by (Holland, 1962).
- Nearest Neighbour: classification of each record is accomplished by a combination of the classes of the k number of records with most similarity historically. Also known as 'k-NN' method and is often used in ECG pattern recognition since its operation relies on statistical pattern recognition. It is a 'supervised' learning algorithm whose results of new instance query are classified according to majority of k-nearest neighbour categories. Its operation is fairly simple: with a query point, it finds k number of objects as training points that are closest to the query point. The classification uses majority vote among the classification of these k objects. Neighbourhood classification is used as the prediction value of the new query instance.

- Rule induction: the simplest method to implement as it only relies on a set of 'if' and 'then' rules derived through observation.

Although the importance of data mining in electronic patient record systems is well understood, we have not addressed solutions for supporting extraction of medical images and structural information. Most image search is currently done by associated text. For example, by adding text markers to accompany an image; search for a medical image then requires prior input of accompanying text and a systematic label system is necessary. Current technologies in image feature extraction are still in a primitive stage. Algorithms used in video understanding for consumer electronics mainly rely on certain image attributes such as colour, contrast and texture. None of these provide adequate solutions for medical images.

4.4   Knowledge Management for Clinical Applications

Electronic patient records are kept in many countries for purposes ranging from patient care to statistical analysis of health risks as well as insurance claims. The behaviour of how data is sought, outlined in (Dawes, 2003), suggests that text search through a vast amount of materials remains a popular practice among physicians. From this observation, we need to find a way to efficiently handle the filing and storage of medical information since it would involve far more data than patients' information alone. Knowledge-based clinical applications span across areas from administration to medical practising and dispensary. The block diagram in Figure 4.24 shows the complexity behind an electronic clinical knowledge system where many entities have their own information to process and share. As we can see, a lot of information is exchanged in this system, both in terms of the types of data and amounts of data. Let us take a closer look at the role of a general practitioner in this context by referring to Figure 4.25,

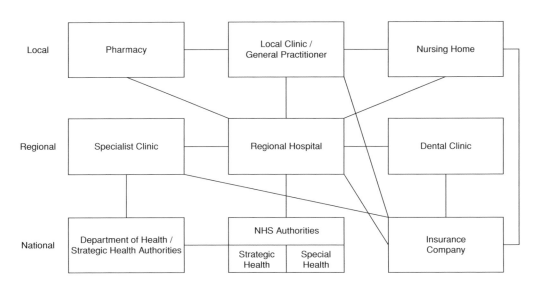

Figure 4.24   Clinical knowledge system

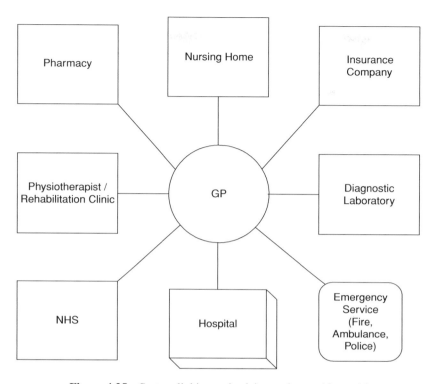

Figure 4.25   System linking a physician to the outside world

which shows the information or knowledge shared with the outside world. There are a lot of interactions between the local doctor and related entities. So, Figure 4.25 is external to the GP's clinic. We then go deeper and expand within the doctor's clinic where we assume that it is a small rural clinic with only one physician. The doctor obtains and shares information in many ways as shown in Figure 4.26 which shows the information dealt with internally inside the clinic. Even within a small local clinic there are many sources of information where knowledge can be obtained from.

Knowledge management is all about creation, transfer, and identification of useful information. The knowledge conversion process is a continually changing and improving process that consists of preservation and enhancement of knowledge. The knowledge conversion process can also be perceived as knowledge creation, transferring and sharing, with the objective of improving knowledge access. The output of the process can be fed back to the input for the next round process for the purpose of continual improvement. In the clinical environment, knowledge management activities are mainly for the creation and maintenance of processes for improving healthcare services so that the general public will be healthier and live longer with less demand for medical services. So, the diagnosis process for providing the best possible treatment would vastly depend on the effectiveness of knowledge management. Following the process in Figure 4.27, constant monitoring of results from previous treatments from electronic patient records would result in more optimized treatment to be developed through

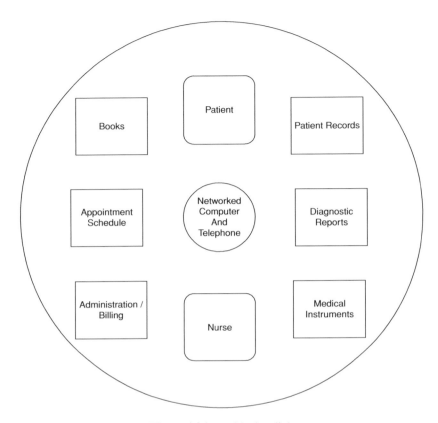

Figure 4.26   Inside the clinic

prior experience. Diagnosis given the symptoms is usually completed according to clinical investigation and laboratory test results. Sometimes diagnostic tests can be time consuming given the urgency of deriving a treatment plan. So, knowledge from previous cases can be of significant assistance in drafting an action plan to provide necessary treatment with minimal delay.

To illustrate this, we take a quick look at the case of using ultrasound to burn off a cancer tumour. A beam of ultrasound when focused on the tumour can rapidly heat it up to temperatures in excess of 70 °C, this process is very effective in damaging cancerous cells as it causes hypoxia that cuts off its oxygen supply. However, there are many restrictions that prohibit its effectiveness to many areas of a human body due to the risk of burning skin and fat. By keeping a record of the effectiveness and result of each treatment, it is possible to compile a list of tumour type and size that such treatment can be used. This example shows us the importance of maintaining a knowledge database for information sharing.

The electronic patient record contains different types of information about a patient in various components, including diagnosis, prescription, appointment record and description of symptoms and treatment provided. They are also useful in keeping a record of archival values when treating future cases with similarities. There are risks and challenges that encumber the

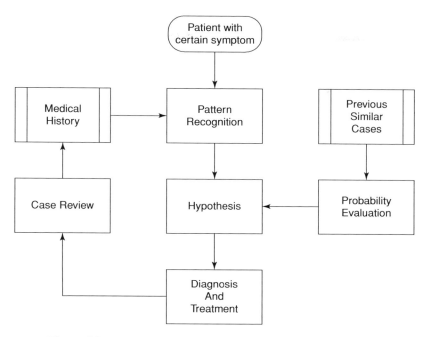

Figure 4.27   Knowledge management for electronic patient records

development of comprehensive electronic patient record systems. First, we refer to the famous 'SOAP' note (Schimelpfenig, 2006) that refers to:

- Subjective: condition of the patient, what symptoms have been described.
- Objective: collection of vital signs, visual inspection to look for signs of abnormalities, and to conduct appropriate laboratory diagnostic tests.
- Assessment: summarize the above based on symptoms and diagnosis.
- Plan: derive an action plan for treatment, e.g. prescription and any follow up action.

Ultimately, this is to facilitate effective patient assessment through knowledge management thereby providing a basis for communication between the patient and healthcare providers. The SOAP note format is often used to standardize healthcare evaluation entries made in clinical records for consistency. As a final note, medical records are legal documents. Therefore, data entry must be done in an accurate and responsible way, and access to information must be strictly controlled.

4.5   Electronic Drug Store

We conclude this chapter by looking at medical information sharing with an electronic drug store for healthcare professionals and end users. The word *telemedicine* may be closely linked to remote dispensary of medicine. Technology makes electronic drug store far more capable than organizing medication for rural areas or people with mobility problems. Although medicine must be physically delivered by some means of transportation, it does provide rural areas better access to medications and related information. One of the features of electronic drug store is to assist with dispensing of medication securely with automated auditing procedures for quality assurance and to reduce administrative costs. It is not simply a vending machine selling non-prescription medicine. Note, incidentally, that the term 'dispensary' is not used here because a dispensary in the United States is an agency that provides substances such as alcohol and legalized cannabis for herbal therapy (Martinez, 2000). Another major application is analysis of drugs so that effectiveness and any side effects can be duly recorded in an efficient and organized manner in the process of developing new medication. Most importantly, electronic drug store serves as a library for patients to learn about proper use of medications, side effects, and automatically generate reminders for replenishment and disposal of expired/outdated medications. It keeps patients connected to their local pharmacies. Pharmacists therefore provide enhanced healthcare service rather than just dispensing medications. Both patients and pharmacists can obtain information about possible adverse drug reactions and allergies. Also, any product recall exercise and expiration of drugs or their respective licence/registration can keep pharmacists up to date.

To the general public, e-prescribing ensures they get the best medicine and the risk of drug mix-up is kept to an absolute minimum as technology is available at every step to ensure proper procedures are followed. Electronic patient records are also integrated to ensure that what they have taken is recorded. In addition, electronic link between physicians, pharmacist, and the patient. Patients can collect their medicine after a doctor visit without the need of bringing a prescription form from the clinic since the pharmacy can retrieve this electronically. The idea of an electronic drug store is to employ a remote drug ordering system such that licensed pharmacists can receive e-dispensing orders and patient records irrespective of the time of day. Pharmacists then check and profile the accuracy of each order and authorise the hospital pharmacy system to dispense the medicine. The pharmacists also monitor allergies, drug interactions, correct dosage and each patient's pharmaceutical history before issuing an authorisation. Also, the system can check if the medication is covered by the patient's insurance so billing can be made accordingly. The generation and storage process of prescription records are all done automatically. Last but not least, patients can be reminded to take medicine at the right time and to replenish their medicine. For the elderly and patients with visual impairments who require long term medication, a small device can be installed at the patient's home as a 'personal medication assistant'. This device, shown in Figure 4.28, can also be implemented as a software application installed on any home or laptop computer with an RFID reader. It reads the RFID tag attached to each bag containing the drug, information about dosage and time for medication, which is stored in the tag so the correct type of medication taken at the right time can be assured. A log of which drug was taken at what time is kept.

Telemedicine and electronic drug store not only help patients get their medications easily and virtually risk free, but also help keep the inventory up to date at all times. This is particularly important when keeping stock for defending against spreading pandemics, an example of such is the March 2009 spread of a new strain of A(H1N1) influenza virus. The stock level of seasonal flu vaccine and related drugs must be closely monitored to ensure that at least the most high risk population gets adequate supply. With a communication link between manufacturers and pharmacies, it can serve as a clinical drug order processing system that can initiate the movement of stock to high demand areas swiftly when the rate of dispatch becomes abnormally high in certain areas.

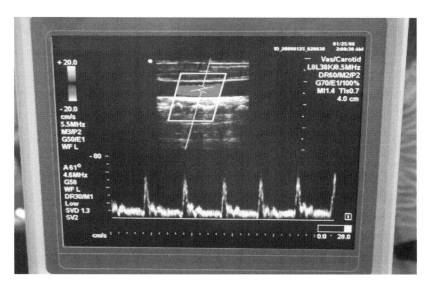

Figure 4.28 Patient monitor

References

Cover, T. M. and Thomas, J. A. (2006), Elements of Information Theory, 2/e, *Wiley Series in Telecommunications and Signal Processing*, Wiley-Interscience USA, ISBN: 978-0471241959.

Cranston, W. I., Hellon, R. F. and Mitchell, D. (1975), Proceedings: Fever and brain prostaglandin release, *Journal of Physiology*, 248(1):27P–29P.

Dawes, M. and Sampson, U. (2003), Knowledge management in clinical practice: a systematic review of information seeking behavior in physicians, *International Journal of Medical Informatics*, 71(1):9–15.

Elmaghraby, A. S., Kantardzic, M. M., and Wachowiak, M. P. (2006), Chapter 16: Data Mining from Multimedia Patient Records, *Data Mining and Knowledge Discovery Approaches Based on Rule Induction Techniques*, edited by Triantaphyllou, E. and Felici G., Springer Massive Computing Series, Germany, pp. 551–595.

Holland, J. H. (1962), Outline for a logical theory of adaptive systems, *Journal of the ACM*, 9(3):279–314.

Koeningsberger, D. C. and Prins, R. (1988), *X-ray Absorption: Principles, Application, Technique of EXAFS, SEXAFS and XANES*, Wiley: New York.

Londeree, B. R. and Moeschberger, M. L. (1982), Effects of age and other factors on maximal heart rate, *Research Quarterly for Exercise and Sport*, 53(4):297–304.

Mackowiak, P. A., Wasserman, S. S. and Levine, M. M. (1992), A critical appraisal of 98.6 degrees F, the upper limit of the normal body temperature, and other legacies of Carl Reinhold August Wunderlich, *Journal of the American Medical Association*, 268(12):1578–1580.

Maintz, J. B. A. and Viergever, M. A. (1998), A survey of medical image registration, *Medical Image Analysis*, Elsevier B. V., 2(1):1–36.

Malik, M. (1996), Standards of Measurement, Physiological Interpretation, and Clinical Use, *Circulation*, 93:1043–1065.

Marchiando R. J. and Elston, M. P. (2003), Automated ambulatory blood pressure monitoring: clinical utility in the family practice setting, *American Family Physician*, pp. 2343-2352.

Martinez, M. and Podrebarar, F. (2000), *The New Prescription: Marijuana as Medicine*, 2/e, Quick American Archives, ISBN: 978-0932551351.

Pickering, T. G. (1999), 24 hour ambulatory blood pressure monitoring: is it necessary to establish a diagnosis before instituting treatment of hypertension? *Journal of Clinical Hypertension (Greenwich)*, 1(1):33–40.

RSNA (2009), Radiation Exposure in X-ray Examinations, January 16, 2009: http://www.radiologyinfo.org//en/pdf/sfty_xray.pdf

Sandsunda, M., Gevinga, I. H. and Reinertsena, R. E. (2004), Body temperature measurements in the clinic; evaluation of practice in a Norwegian hospital, *International Thermal Physiology Symposium: Physiology and Pharmacology of Temperature Regulation*, 29(7):877–880.

Schimelpfenig, T. (2006), *NOLS Wilderness Medicine*, Stackpole Books USA, ISBN: 0-8117-3306-8.

Shannon, C. E. (1948), A mathematical theory of communication, *Bell System Technical Journal*, 27:379–423.

Tempkin, B. B. (2009), *Ultrasound Scanning: Principles and Protocols*, 3/e, Saunders.

Tobin, M. (2001), Effects of Lossless and Lossy Image Compression and Decompression on Archival Image Quality in a Bone Radiograph and an Abdominal CT Scan, an online tutorial: http://www.mikety.net/Articles/ImageComp/ImageComp.html

5

Wireless Telemedicine System Deployment

As we have seen in the previous chapter, data conveying information about a person's health can be obtained from many sources. Different types of data have different ways of being acquired and many have different requirements on transmission and subsequent processing. We have learned how various types of medical data can be captured and what to consider when making the data suitable for transmission through telemedicine networks. Vital signs and medical images are different in many ways; some have more stringent requirements than others. The diversity of data acquisition makes both instantaneous and long-term measurement necessary in order to cater for different health monitoring situations. One key requirement in common is an efficient and reliable communication network to support patient caring. Network implementations are determined by the specific application supported so that they are designed to satisfy the specific requirements imposed by the type of data sent. For example, sending an X-ray radiograph has very different requirements in terms of bandwidth than sending a prescription form that contains plain text information.

Any communication channel has a specific theoretical limit to the amount of data that can be conveyed, this applies to everything wired or wireless. The channel bandwidth governs how many data bits can 'pass through' the given channel in one second. A network must therefore make use of communication channels that are capable of delivering all the data for an application without overflow ('flooding' will happen if too many data bits attempt to get into the channel at a rate too fast for the channel). To understand more about its importance, we take a look at an example of attempting to send a Full-HD video clip over an analogue telephone channel, whose channel bandwidth is 3 100 Hz. Obviously, we can tell straight away that far too many data bits are there for the available bandwidth without even doing any calculation. Even with *data compression* we still need bandwidth in the magnitude of MHz for HD video transmission.

In digital communications, information is acquired either as a *block* or *stream*. The bursty nature of information, in the case of randomly taking a one-off measurement, does not usually have any statistical pattern of occurrence. So, a discrete block of data is collected when each reading is taken. No more data will follow until the next set of readings comes. An example of this kind of random event is the hospital A&E admission where sometimes there is no patient

Telemedicine Technologies: Information Technologies in Medicine and Telehealth   Bernard Fong, A.C.M. Fong, and C.K. Li
© 2011 John Wiley & Sons, Ltd

while at other times the department may be treating several patients at the same time. The discrete probability distribution of data acquisition means statistical analysis of information flow is best dealt with using Poisson distribution modelling (Shmueli, 2005). In contrast, continuous monitoring, such as in the case of collecting data from a wearable device for health monitoring, will generate a stream of data as information will come in incessantly at a certain rate. We therefore handle sequential data of infinite duration, that is, until monitoring is interrupted. Audio and video information is usually of such a nature. To understand more about how a communication system handles data of discrete blocks and continuous stream nature, we go back to the earlier example of attempting to send a video clip over a telephone channel of bandwidth 3 100 Hz originally designed only to carry mono voice signal. If the video comes as a burst of, say five seconds, short clip; and nothing follows for the next few minutes. The entire clip can still get through with lengthy delay. The channel has sufficient time to 'swallow' the large amount of data, just like pouring water into a narrow pipe through a funnel. If the funnel is large enough to act as a 'buffer', and the water stops coming in before the funnel overflows; it is still possible to get the water through without spilling over. However, a continuous flow of water will not get through as in the case of a stream. Imagine what would happen if we leave the water tap fully open and let it run down a funnel continuously into a narrow pipe that does not have sufficient capacity to carry the water flow. The result is obviously overflowing from the funnel and water spills over. Exactly the same would happen in data communications if we attempt to throw too many data bits into a channel that does not have adequate bandwidth to carry the data.

Communication networks are essential parts of modern healthcare systems that play a vital role for information exchange. With the capability of supporting a vast range of medical services as seen in previous chapters, networks are inherently developed to support a multiplicity of innovative services as technology advances. In this chapter, we shall learn how to beat the challenges of developing and maintaining a future-proof network that will incorporate new features as and when they become available. From the fundamental knowledge on digital communications gained in Chapter 2, we proceed by first looking at some theories behind network planning and exploitation followed by necessary measures to ensure the network can be expanded for the future. As many networks are built on existing frameworks, what needs to be considered will be discussed. We then explore the pros and cons of outsourcing, and conclude this chapter by exploring network quality assurance.

5.1   Planning and Deployment Considerations

To thoroughly understand what lies behind network planning for telemedicine, we must first get a good understanding about what goes on behind the scene. A good starting point is by referring to a primitive computer network that consists of two PCs linked together in a *peer-to-peer* (P2P) structure in Figure 5.1. Each PC has a network interface card (NIC) that connects it to the outside world, which is another PC in this example. As far as the PC is concerned, whether it is connected using a cable or via a wireless link makes no difference as long as data can be reliably exchanged between the PCs. The key feature of a P2P network is that it does not have a centralized location so that all nodes (members of the network) are of equal status heretically. Before advancing into the technical details of communication networks, let us first look at what happens inside the PC in the context of data communications by introducing the Open Systems Interconnection (OSI) reference model (ITU-R X.200, 1994).

Communication Channel
(can be either a cable or a wireless link)

Figure 5.1   A simple peer-to-peer network (most basic form of a network)

5.1.1   The OSI Model

The OSI reference model provides an outline for network communications and its main purpose is to serve as a guide for network design. It is essentially a descriptive model for *layered* communication, i.e. network architecture is split into different layers each specifying a set of functions in data processing and formatting. The standard model in Figure 5.2 consists of seven distinctive layers. Each is responsible for a set of tasks. Communication between any two adjacent layers is said to be 'direct'. This involves exchange of data blocks through a port

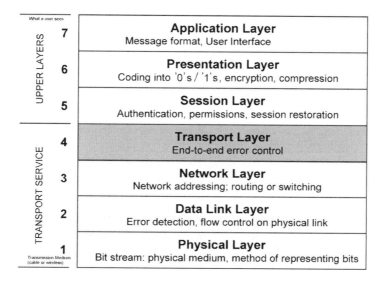

Figure 5.2   The seven-layer OSI model

of a layer known as a *service access port* (SAP). Of the seven layers, they are broadly classified into two groups, namely the upper three layers known as *host* layers, and the bottom three layers grouped as *transport* layers. The middle layer lies in a grey area that some literatures group into *host* layers while others group into the *transport* layers (e.g. the definition given in *wiki*). As the middle layer itself is named 'Transport', it would logically be more appropriate to be considered as belonging to the lower transport layers.

The OSI model basically classifies the entire communication process into functional layers. Each layer of a given host, e.g. PC A of Figure 5.1, a communication process is maintained with a peer process on the corresponding layer of PC B. Put simply, any given layer n of A talks to B's layer n, where n can be any of the seven layers of the OSI model. The processes carried out at layer n are collectively *layer n entities*. Since no direct connection link is established between layer n of the two PCs, communication between the two layers n on both sides is said to be 'virtual'. Communication is accomplished by the exchange of *protocol data units* (PDUs). A PDU is essentially an 'envelope' that carries the data inside it. The user data carried inside a PDU is known as a *service data unit* (SDU). So, an SDU is part of a PDU that also contains a *header* of information about the data but is not part of the user data. The function of each layer n entity is managed by a set of rules called *layer n protocol*. Each layer n entity functions by using control information to construct a header, along with the SDU to produce a PDU that is sent down to the layer $(n - 1)$ below for further processing. The function of layer n usually entails reception of PDUs from the layer above $(n + 1)$, delivering it to its peer processes then passing it down the stack (i.e. to the next layer below) and eventually to manipulate the data into a suitable form for transmission through the communication channel.

In case the data block exceeds the maximum size that can be handled by the next layer $(n - 1)$, the 'segmentation and assembly' process becomes necessary to break the data block down into smaller units that can be 'swallowed' by layer $(n - 1)$ and subsequently put the segmented data back on the receiving end. Here, the SDU is segmented into a multiple number of layers n PDUs. These smaller units are then passed further down the stack until it is sent out from PC A and reaches PC B. Reassembly (joining smaller pieces back together, as the reverse process of segmentation) is then performed at PC B's layer n.

Next, we take a brief look at the functions of each layer shown in Figure 5.2, starting from the top:

- *Application Layer*
  The top or the 7th layer, where human-computer interface (HCI) is supported, makes direct interaction between a user and the application software possible. For example, a doctor generating a prescription does so by entering the information about the patient and the drugs, and sending it off to the pharmacy nominated by the patient. This entire process, as seen by the doctor, is handled by the application layer. Applications such as database entry, word processing, web browsing, email, etc., are all handled by the application layer.
- *Presentation Layer*
  Context between application layer entities are established here. It supports the application layer on data representation. Its main function is to convert information from human users generated with application software into a form suitable for the computer to process. So, data is 'mapped' into Session Protocol Data Units and passed down to the next layer. Different computer types running different operating systems (OS), e.g. communication between a PC and a PDA, may use different code sets for information representation and

this is the presentation layer's task to convert data into a 'machine-independent' format for transmission.

- *Session Layer*

  This is where connections between nodes (computers) are administered. It establishes, manages and terminates connections. This is also where the mode of communication, e.g. simplex or duplex, is controlled. Synchronization that may be required by some error recovery services is supported by the session layer. This is particularly useful for handling a long data stream such as transmitting ECG data.

- *Transport Layer*

  Reliable data transfer between end users is assured by the transport layer. The link reliability is managed through flow control, segmentation and reassembly, as well as error control. It uses the services provided by the underlying network for imparting the session layer with the necessary quality of service (QoS) for data transfer; and keeps tracks of the segments and initiates re-transmission when necessary.

- *Network Layer*

  Here, data is converted into 'packets' for transmission over the network. The process of identifying the optimal path for each packet to reach its destination, called *routing*, is carried out by passing through a number of transmission links (in almost all telemedicine networks, far more network nodes and transmission links are present than the simple example shown in Figure 5.1). Effective routing requires information about the links' conditions from other network nodes. So, congestion in certain parts of the network can be isolated when an excessive amount of data is exists.

- *Data Link Layer*

  Data transfer between network entities and error detection/correction are all taken care of here. Data blocks are converted into 'frames' where a header that contains control, address information, check bits for error detection; and framing information to mark the boundaries of each frame are inserted. Also in the data link layer, medium access control (MAC), manages data transmission from nodes into a communication channel (the 'medium'), hence the process of controlling the access to the medium.

- *Physical Layer*

  Finally, right at the bottom of the stack we reach the first layer. The electrical and physical specifications between the node and the medium are defined here. These are attributes like how the '0's and '1's are represented; what the data rate, hence signal duration, is; pin configurations of plugs and sockets used are also defined here so that what the data from a specific pin represents can be known. Since the physical layer only deals with sending binary bits through a communication medium, it does not concern itself with the actual meaning of the stream of data bits. So, whether the data contains medical images or body vital signs it would make absolutely no difference to how the physical layer handles the data. Modulation is also performed here. Establishment and release of physical connection is another main task of the physical layer.

5.1.2   Site Survey

Surveying is a very important step at the early stage of network planning. This is vital to establish the correct number and placement of radio stations or access points in any wireless networks. Establishing the impact on signal reception under different scenarios cannot be taken lightly, site surveys and simulations allow 'stress tests' to be carried out to find out problem areas such as poor reception, interference, susceptibility to hacking, etc. (Hummel, 2007). Unlike the stress test performed by banks in May 2009 that attempted to simulate the impacts on business operations but ended up giving virtually no hints on their true states, there are so many different possibilities that can cause temporary interruption of a link or corrupt a large block of data that cannot be taken lightly; test on wireless network sites would reveal far more useful information about 'what if'. There may be questions like 'what if moving objects randomly obstruct a radio link that requires direct line-of-sight?', 'what if frequent heavy downpour degrades a radio path?', or 'what if coverage needs to be expanded to serve the new building across the road next year?' As we can see, there are all kinds of questions that we may need to address when we start planning a new network. Quite simply, we need to seek information about the location of each radio station and its coverage area.

Site survey entails measuring wireless signals and using the measured data for network planning and optimization. The major factors to study are coverage, capacity, interference, and physical obstacles. Coverage area is directly related to the signal strength. To illustrate this, we recall the procedures of connecting our laptop computer to a wireless network in a café. When the network is found we see a number of bars of ascending heights, the more of these bars we see the stronger the signal strength we can pick up from the access point. As we move further away from the access point, these bars vanish one after another until we move to a certain location where all bars eventually disappear and the computer is disconnected from the network. This is where network coverage is no longer available. The loss of signal strength (attenuation) as we move away from the access point, follows the inverse-square law, such that the signal strength $S(d)$ is governed by the distance d between the access point and the receiver such that:

$$S(d) \propto \frac{1}{d^2} \qquad\qquad\qquad (5.1)$$

This is under the assumption that we maintain a direct line-of-sight (LOS) path from the access point without any obstacle in between.

Network capacity concerns the maximum number of users that can be connected to the network at any given time. It also governs how much data can be transferred simultaneously by all users. Therefore, wireless networks can be saturated when its maximum capacity is taken up. When this happens, any other users that attempt to make a connection will either be blocked (denied from access), or all users will experience a degradation in QoS (data transfer becomes slow and intermittent network outage becomes more frequent).

Interference describes the strength of all distracting signals that come within range of coverage. This can be a major issue when the network operates in a frequency band that is shared by other systems. For example, an IEEE 802.11n WLAN operating at 2.4 GHz may receive interference from a nearby coreless telephone that also uses the same carrier frequency. Also, WLANs of close proximity can interfere with each other, as in the case of an apartment block where several units have their wireless routers operating at the same time, at the same carrier frequency with the same channel. Connection reliability and data speed can be jeopardized due to this kind of interference.

Physical obstacles can cause all kinds of problems to wireless signals as discussed earlier in section 2.4. Absorption and reflection can be serious problems in many situations. Walls and partitions, especially thick concrete with steel beams, can shield off radio waves. Also, glass panels with film coatings or embedded wire mesh can degrade propagating signals.

During a site survey, measurement of signal strength and interference is conducted in various locations and usually a laptop computer installed with a network management system (NMS) that captures the measured data. Test access points are sometimes used and moved across various locations to test relative signal reception quality from a reference location in order to find out the optimal location to install an access point. Some modern surveying software allows prior input of the location map so that the surveyor can simply stroll along the site with a laptop computer while it measures and analyzes the received signal from each access point. In large sites such as a hospital, it may be practically impossible to scan through the entire site so estimation can be based on extrapolation from certain sampled areas. Site survey can give an indication on what to expect but readers should understand that the results do not imply the completed network will behave as test results since there are far too many uncontrollable variables involved. As a final note, certain operational and safety limitations may be imposed by local authorities. Any such applicable requirements must be met before network setup is completed. In addition, access point placement must be free from interference from any nearby delicate medical instruments.

5.1.3   Standalone Ad Hoc Versus Centrally Co-ordinated Networks

Wireless networks can be deployed with either standalone access points or centrally co-ordinated. The former, such as that depicted in Figure 5.3, utilizes integrated functionality of

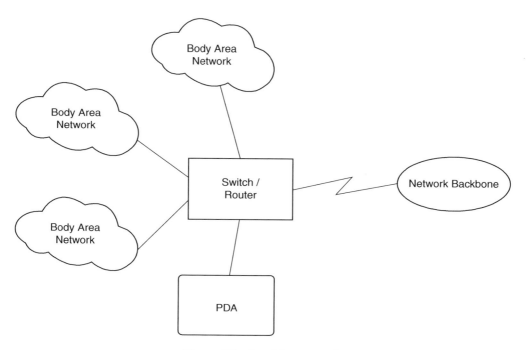

Figure 5.3   An ad-hoc network

each access point to enable wireless network services. Each access point in the network operates independently of each other and each access point is configured separately so that it does not respond to changing network conditions such as data traffic congestion or neighbouring access point failure. In this network configuration, it does not have any centralized location that deals with user access or data flow control. Each body area network operates quite independently from each other. For patient monitoring applications that usually involve continuous streaming of data, these require high energy to maintain adequate performance (Zou, 2005). Such simple arrangement without centralized control fails to deal with issues such as power management, packet losses, security attacks and similarities in applications that lead network performance degradation.

In a centrally co-ordinated wireless network, each access point is relieved from most of the data management tasks as they are taken care of by a central controller. Network performance monitoring throughout the entire network can be done centrally. Expanding coverage area with this kind of network arrangement can be easily accomplished by plugging more access points into the controller and letting it monitor traffic flow among all access points. The controller can be programmed to reconfigure each access point independently due to change in network conditions, for example, disabling failed access points or re-routing traffic for load balancing. This provides mechanisms for a 'self-healing' network. Its configuration is very similar to that of Figure 5.3, except that the switch or router that connects different access points or body area networks together needs to be replaced by a controller.

Standalone configuration is usually more suited for smaller isolated wireless coverage areas with a very small number of access points, or in situations where temporary extended coverage is needed to serve a nearby area. Otherwise, centrally co-ordinated wireless network configuration is more desirable since it facilitates ease of deployment and rapidly responds to real-time changes in network conditions.

5.1.4   Link Budget Evaluation

This is an essential step to determine the link range or coverage by giving the system an operating margin in case any unforeseen events degrade the operating environment. For outdoor networks where rain can severely impact signal propagation, other factors such as modulation schemes and polarization also have a significant impact on the radio link (Fong, 2003a). The link budget describes the gains and losses incurred throughout the entire communication system of Figure 2.2. The concept is very simple: the received signal power P_R, after going through the entire communication system of antennas, channel (air, physical obstacles along the propagating path, etc.), and all cables/wires that connect components such as that between the receiving antenna and the demodulator; can be described as:

$$P_R = P_T + G - L \qquad (5.2)$$

where P_T is the transmitted signal power in dBm, G and L are the sums of all gains and losses in dB throughout the entire communication system, respectively. Let us go back to Equation 5.2 and expand on the link coverage, radio wave spreads as it propagates along the path. Doubling the propagating distance will result in the received power reducing to a quarter so that:

$$L = 2 - \log_{10}\left(\frac{4\pi d}{\lambda}\right) \tag{5.3}$$

(NIST, 2006) provides a link budget calculator for a rough estimation of link budget in an outdoor environment. Since the transmitting and receiving properties of any pair of transceivers can be quite different, it is usually necessary to calculate the link budget for both signal directions.

In general, a telemedicine network requires at least 10 dB of link margin to tackle variations of signal strength due to reflection. Further, an additional 30 dB is necessary in case of polarity mismatch between antennas in an orthogonally polarized configuration. So, link margin is one important parameter that distinguishes telemedicine from general purpose wireless networks. We need to ensure that the network remains reliable even under persistent heavy downpour. The amount of extra link margin that we can afford to provide depends primarily on transmitter design and site environment. Maximizing the affordable link margin would ensure optimal system reliability.

5.1.5   Antenna Placement

Where to place an antenna is sometimes a tedious question since, in practice, the location offering optimal performance is not a feasible location for placement. Also, impedance matching for the efficient transfer of energy between the antenna, radio, and cable connecting between the antenna and the radio have to be identical in order to avoid loss due to impedance mismatch. Since most antennas inherent impedance that differs from that of the connecting cable, impedance matching circuitry is usually necessary for transforming the antenna impedance to that of the cable. Impedance match is measured by the *Voltage Standing Wave Ratio (VSWR)*. The VSWR should be less than 2:1 in order to ensure that a vast proportion of the power is forwarded with minimal reflected power. A high VSWR indicates that the signal is reflected and lost.

The ratio of VSWR and reflected power defines an antenna's bandwidth. The *percent bandwidth* is constant relative to the carrier frequency f_c expressed as:

$$BW = \frac{f_H - f_L}{f_c} x100\% \tag{5.4}$$

$$f_c = \frac{f_H + f_L}{2} \tag{5.5}$$

where f_H and f_L are the highest and lowest frequency in the band, respectively.

The angle of coverage or directivity of an antenna needs to be considered in antenna placement since many antennas do not provide a 360° degree omni-directional coverage. It should be noted that omni-directional coverage usually refers only to horizontal directions but does not necessarily extend to the vertical or elevation plane. The directivity of an antenna describes its focus of energy in a particular direction when transmitting or receiving energy from a transmitting antenna pointing towards it. Essentially, the directivity is measured as the ratio of an antenna's efficiency versus gain. Although most wireless routers we use at home are equipped with a cylindrical monopole antenna for $360°$ degree coverage, there are many antennas that are highly focused with beamwidth of only a few degrees. Narrow beamwidth antennas are used when longer coverage is necessary by focusing energy towards a certain direction. This situation is very similar to comparing an ordinary light bulb with a spotlight of the same power rating, where a spotlight concentrates its illumination over a more confined area but is much brighter than a light bulb of the same wattage covering the same area. The radiation pattern, showing the relative strength of the radiated field in different directions from the antenna, in the coverage area close to the antenna diverges from that of the pattern over long distances. This leads to the introduction of the terms 'near-field' and 'far-field', also known as induction and radiation field, respectively. Far-field is normally used for measuring an antenna's radiated power. The minimum distance d to conduct far-field measurement is given by:

$$d = \frac{2.l^2}{\lambda} \quad\quad\quad (5.6)$$

where l is the length of the longest side of any dimension of the antenna and λ is the wavelength of the carrier frequency as in Equation 2.6. Near-field measurement is far less important for placement consideration of antenna. The only situation where it is useful is when deducing the minimum safe distance, in which dealing with ultra-high power antennas may radiate sufficiently large amounts of energy to cause health concerns.

5.2   Scalability to Support Future Growth

Network scalability refers to the ability of a network to scale up for increase in capability, in terms of performance, capacity and coverage. Any communication network should be designed to handle future expansion in terms of both data throughput for new services and number of subscribers. It should also be made possible for extending coverage area. A properly designed scalable network should at least maintain its performance, if not enhance it, when expanded.

Scalability often involves installation of new hardware to an existing network. While a network is fully operational any interruption to its normal operation should be restrained during the process of expansion. Maintaining network availability when works are carried out is particularly important with a system that supports life-critical missions. It is unreasonable to assume that a network can be temporarily shut down for improvement work to be taken since no mechanism exists for predicting when a telemedicine network is not used, since accidents won't wait until service resumption to happen. Scalability almost certainly involves laying new cables in the case of wired networks, whereas with wireless networks several parameters can be adjusted within the network infrastructure. It is therefore far easier to perform enhancement work on a wireless network than with cables.

We begin our technical coverage by referring to (Fong, 2004) as we look at what relevant parameters exist within the network backbone. Obviously, it is impossible to alter anything along the path, namely the air as signal propagates through it. We have to work on the transmitter side and make sure that each receiver in the network is capable of processing the incoming stream of data in the case of increasing data throughput. Remember, the main objective is to efficiently utilize available system resources and minimize errors so as to keep the need of re-transmission to an absolute minimum. Before we look at how to fiddle around with various system parameters, let us recap on what within a wireless network can be altered to improve network capacity.

5.2.1   Modulation

Modulation, being the process of 'putting' data bits into the signal, can be changed to 'squeeze' more data bits into the signal. We can see this in the *constellation diagram* of Figure 5.4, which shows a representation of the signal quality as well as any presence of distortion. As seen from

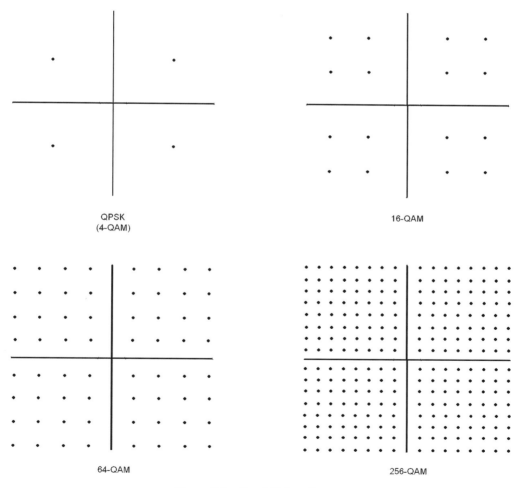

Figure 5.4   Constellation diagram

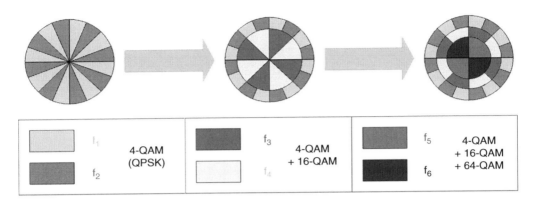

Figure 5.5   Coverage enhancement through augmentation. Reproduced by permission of © 2002 IEEE

the constellation diagram whose axes represent the signal's amplitude and phase, higher order modulation (e.g. 64 or above) have more 'dots' so that more data is carried. At the same time, an increase in order packs the dots closer together while squeezing more bits into the signal. The nearer these dots get towards each other, identifying each individual dot will become more difficult. The net result is a trade-off between spectral utilization efficiency and receiver structure complexity since more complicated receivers are required to resolve each dot on a closely packed constellation diagram. With something as high as, say 1 024, the amplitude and phase difference between adjacent signal points may make them indistinguishable from each other.

In general, signals transmitted with lower modulation order, e.g. QPSK, can be properly received from a greater distance with an identical transmission power compared to higher order modulations as signal loss is a less significant issue. It is therefore possible to combine the coverage distance of QPSK to serve only areas further away from the transmitter, and the higher bandwidth efficiency of 16-QAM for serving receivers closer to the transmitter. Further augmentation is possible to enhance overall coverage, as illustrated in Figure 5.5.

5.2.2   Cellular Configuration

A wireless network can be either single cell or macro cells (a cellular infrastructure with multiple cells to cover the overall coverage area, as illustrated in Figure 5.6). They can both utilize *frequency reuse* technique but in different ways. Frequency reuse is a method of enhancing spectral efficiency and network capacity through reusing channels and frequencies of the same network that is achieved by dividing an RF radiating area (coverage area) into segments of a cell, as shown in Figure 5.7 where a simple example shows the reuse of two different frequencies. These frequencies, although in the same frequency band, need to be allocated with adequate separation from that of all neighbouring segments to minimize any

Hexagonal cells with four different frequencies

Figure 5.6   Cellular coverage. Reproduced by permission of © 2002 IEEE

risk of interference. Any given frequency is reused at least two segments away from each other.

In a single cell structure, broader geographical coverage is usually provided by high gain antennas with direct LOS to receiving antennas and frequency reuse is possible by using different polarizations. Whereas macro cell systems use spatial frequency reuse that can usually provide acceptable signal reception properties without LOS.

So, spectral efficiency can be improved through a combination of higher order modulation and frequency reuse. There is, however, an inherent problem as frequency reuse amplifies *co-channel interference*, which means that two sufficiently nearby channels sharing the same

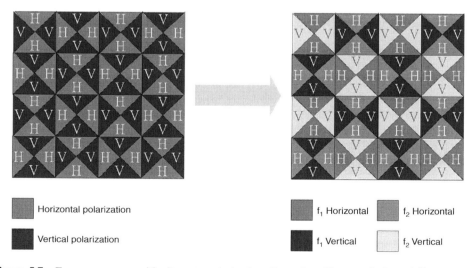

Figure 5.7   Frequency reuse with alternate polarization. Reproduced by permission of © 2002 IEEE

frequency interfere with each other. This is usually caused by problems such as network congestion or bad frequency planning during the system design stage. This, contradictorily, reduces the modulation order. The spectral efficiency, measured in bits per second per hertz within a cell (quite simply, bit rate per Hz of bandwidth per cell = bps/Hz/Cell, or BHC for short), is the data rate sent across each cell per Hz of the channel bandwidth.

Co-channel interference in a single cell configuration is primarily caused by scattering from reuse sectors within the cell. Since macro cell configuration employs frequency reuse in spatially separated segments, and the sharing of frequency amongst nearby segments induces co-channel interference. By applying Shannon's Theorem (see Chapter 2), frequency reuse is possible if:

$$BHC \leq \frac{\log(1 + C/I)L}{K.m}. \tag{5.7}$$

where C/I is the channel to interference ratio, L is the number of times a channel is reused and K is the spatial reuse factor ($K = 1$ in single cell configuration); and m is the overhead assigned to *guard bands*. A guard band is defined as a narrow portion of frequency band that is assigned to separate between two channels of similar frequencies without carrying any data, as illustrated in Figure 5.8.

Looking at Equation 5.7 more closely, we can deduce that frequency reuse can be optimized by increasing L in a single cell configuration or reducing K with macro cell. In practice, C/I for single cell and macro cell structure can be approximated as Equation 5.8 and 5.9, respectively (Sheikh, 1999):

$$C/I = \frac{c}{L} \tag{5.8}$$

$$C/I = c.K^2 \tag{5.9}$$

where c is an arbitrary constant specific to a given network deployment.

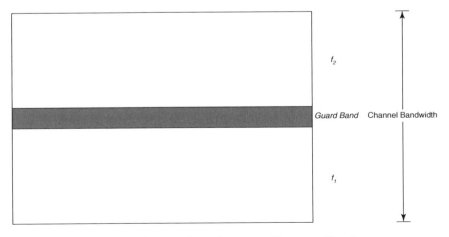

Figure 5.8   Sub-channels separated by a guard band

Frequency

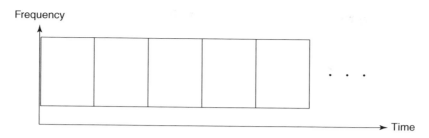

Time

Figure 5.9   Time division multiplexing

5.2.3   Multiple Access

As its name suggests, multiple access refers to multiple devices connected to a single wireless channel so that its available bandwidth is shared by all devices within the network. Rephrasing this sentence tells us that we are simply looking at 'splitting' the channel into different portions using some kind of 'multiplexing' technique. In its simplest form, we can split the channel into either time or frequency slots that lead to *time division multiple access* (TDMA) or *frequency division multiple access* (FDMA). These names may sound familiar as we heard about them many times when digital cellular phones were launched around the early 1990s. The operating principle is very easy to understand if we look at Figures 5.9 and 5.10.

In Figure 5.9, the channel is split into different time slots. The time slots are of equal duration in this particular example but this does not necessarily have to be the case. Each transmitting device is assigned with a time slot and the next device will use the next time slot. Taking an example of three devices, A, B, and C, the duration of each time slot is 10 ms. So, when the transmission process begins at time $= 0$, A will have exclusive access to the channel for the first 10 ms. At time $= 10$ ms B will take over and given exclusive use of the channel for the next 10 ms so that it stops transmitting at time $= 20$ ms. This then allows C to take over the channel between 20 and 30 ms. The entire channel sharing process then repeats itself when C's turn is completed at time $= 30$ ms so it returns to A again for its next time slot of another 10 ms and so on. This, of course, only briefly describes the theoretical principle. In practice, switching between transmitting devices takes a finite amount of time so releasing the channel by A and B starts transmitting is not simultaneous. A short amount of time as overhead for switching must

Frequency

Time

Figure 5.10   Frequency division multiplexing

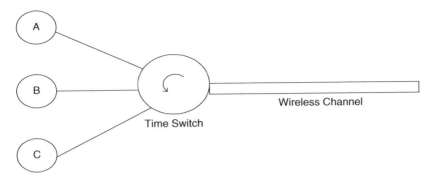

Figure 5.11   Switching between time slots

therefore be taken into consideration. Logically, the process can be described by Figure 5.11, where a logical switch allows each transmitting devices exclusive access to the channel.

Instead of splitting the channel into time slots, FDMA splits the channels into different frequency bands within the allocated bandwidth as shown in Figure 5.12. For example, a channel assigned with the band 100–400 MHz, when equally split between three transmitting devices, will be as follows: three transmitters require subdividing the 300 MHz frequency range (from 100 to 400 MHz) into three equal portions would mean we have 100–200 MHz, 200–300 MHz, and 300–400 MHz sub-channels to be assigned to the transmitters. Each gets 100 MHz or one third of the total channel bandwidth. Again, in practice we'll never be able to allocate a full 100 MHz for each sub-channel simply because the band pass filters (each source is connected to a band pass filter instead of a common switch with TDMA as in Figure 5.11) can never have sharp cut off at one exact frequency but a gradual cut off over a narrow frequency range. We need to spare a guard band for filter cut off in order to leave an adequate margin for the filters. In Figure 5.13, we can see an 'ideal filter' cuts off at exactly one frequency. Of course, ideal filter does not exist in real life. Instead, all practical filters require a range of frequencies to cut off as the cut off process is a gradual process. Better performing filters will cut off with a steeper slope.

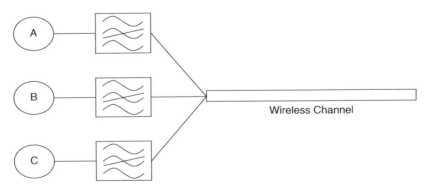

Figure 5.12   Filtering for different sub-bands

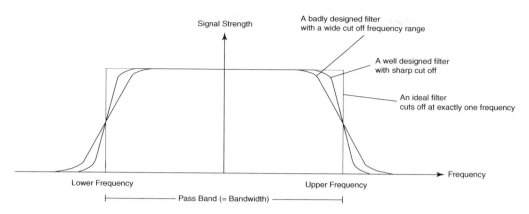

Figure 5.13   'Ideal' filter with sharp cut off

These access techniques have an impact on the way a wireless network is shared between different users and entities. While other multiplexing techniques such as Code Division Multiple Access (CDMA) are also used, TDMA and FDMA are the two most popular options. As we can see the main trade off between the two are full bandwidth available for periods of time versus constant availability of reduced bandwidth. It is therefore most appropriate to utilize TDMA for downstream data traffic and FDMA for upstream due to the bursty nature that makes TDMA a better choice, whereas FDMA provides a constant pipe making it more suitable for upstream data traffic. Dynamic bandwidth allocation improves channel efficiency by assigning more resources to that of higher demand.

5.2.4   Orthogonal Polarization

It is possible to expand data throughput by fiddling around with antennas. For example, with two signal paths of both vertical and horizon polarizations, we can essentially have two separate channels simply by mounting two antennas of orthogonal polarizations perpendicular to each other. Relative to the earth's surface, the electric field of a vertically polarized antenna is perpendicular, whereas for horizontally polarized antenna is parallel. These are both said to be linearly polarized antennas, a classic example would be the old fashioned TV antennas mounted on rooftops similar to that in Figure 5.14. As we can see, this type of TV antenna is parallel to the earth's surface, this is therefore an antenna of horizontal polarization. We can effectively double the number of channels by adding another antenna to make use of another polarization.

Antennas of circular polarization also exist, as shown in Figure 5.15 as an example. Here, the polarization plane rotates in a circular pattern such that it completes one rotation per wavelength. For example, the polarization would have rotated by 360° in one metre if the wavelength is 1 m. Energy is radiated in all directions, including horizontal (0° and 180°), and vertical (90° and 270°) planes. The propagation direction can be either clockwise (right hand circular or RHC) or anti-clockwise (left hand circular or LHC). Generally, circular polarization is more suitable for non-LOS situations when meandering through physical obstacles since

Figure 5.14   Conventional outdoor TV antenna

the reflected signal bounced back to the transmitting antenna upon striking an obstacle will be different from that of the propagating signal.

Another way of achieving scalability with antenna placement is by *sectorization*, where antennas are added on demand by an increased number of sectors as shown in Figure 5.16. In this example, initial deployment is set up with one antenna providing an omni-directional

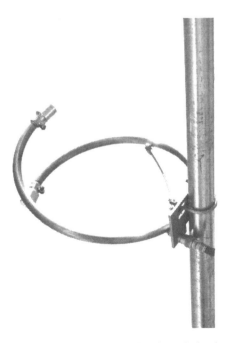

Figure 5.15   Antennas of circular polarization

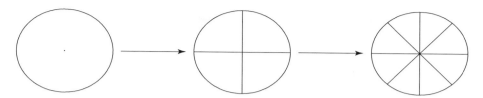

Figure 5.16   Sectorization from one cell into four and eight sectors

coverage. As demand grows, three more antennas are added so that each only serves a 90°
coverage thereby providing three more channels for the same area. Further growth can be
supported by further segmentation, as in this example showing the option of doubling from
four to eight sectors.

5.3   Integration with Existing IT Infrastructure

Many telemedicine networks are built upon some kind of existing network. For example, by
monitoring an asthma sufferer at home (Gibson, 2002) the system might utilize the home
network and its Internet connection via the ISP. What we have here is an addition of the
asthma self-monitoring system to an existing network in Figure 5.17. In this simple illustration,
shutting down the home network temporarily during the installation process probably will not
cause too much inconvenience. Network integration in more complex systems such as an
enterprise network of a hospital would be a far more complicated issue since it is impossible
to expect the network to be shut down for the integration process. As with any maintenance
work, one key pre-requisite is to ensure any work carried out does not affect the operation of
other parts of the network.

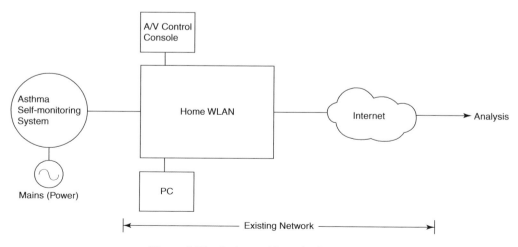

Figure 5.17   Asthma self-monitoring system

To facilitate any network maintenance, the building map depicting physical network is a vital piece of document that details location of all wiring, access nodes, and associated devices. This allows physical integration for connecting all new devices to the right place. Both data network connectivity and power must be connected to the new devices as the vast majority of monitoring devices cannot draw power directly through the data cable.

Logical integration involves configuration of the new portion of an overall network. Different network architectures of the existing network will have different integration requirements. Most modern networks, including all IEEE 802.11 based WLANs, are IP networks that make connections easy with a comprehensive set of standards. Some older networks, however, may have legacy networking protocols that require additional work to be carried out during the integration process.

5.3.1 Middleware

One important and useful piece of tool in network integration is a *middleware*. It is essentially a software bridge that links any systems onto a network by providing necessary services. The prefix 'middle' defines itself as a piece of software that sits in the middle between the applications and the operating system (OS) of a computer. A comprehensive tutorial can be accessed from (Krakowiak, 2009). Middleware provides features such as communication, data access, and resource control for connection of devices with different platforms (Rimassa, 2002). Middleware with a wide range of features is designed for different healthcare applications (Spahni, 1999). The main purpose of the middleware is to facilitate integration of computer systems, medical instrumentation, monitoring systems, databases, etc.

Middleware is frequently configured to access databases as it facilitates communication between applications that often associates with the term Enterprise Application Integration (EAI), describing the consolidation between different applications used throughout various entities within the healthcare system. EAI is designed to address problems such as integration and connectivity, where data integration is usually accomplished by incorporating application programming interfaces (APIs) for communication with legacy systems in order to ensure compatibility. An API provides an interface for a given application to obtain services from a computer's OS or libraries.

5.3.2 Database

A database stores a vast amount of information in the form of *fields*, *records*, and *files* in such a way that the collection of information is organized for easy retrieval as an individual item or a group of items that satisfy certain user-defined selection criteria. A *field* is a single piece of information and a *record* is one complete set of fields, whereas a file is a collection of records. For example, a field may store information about the medication prescribed to a patient during this morning's consultation, and a record contains all information related to the consultation session such as medication prescribed, nature of visit, symptoms, body temperature, etc. and a file is a collection of medical history that belongs to this patient. Database size can vary from a small clinic that may contain several hundred patients' medical histories, each stored in a file; all the way to a national health database which may contain millions of files of patients, separate storage for information about suppliers, manufacturers, pharmacies, etc. It may total

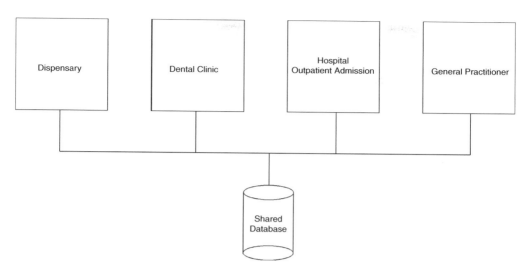

Figure 5.18   Database sharing in a hospital

to billions of files stored together in a logical structure. Information sharing among different databases of varying size and format can be a nightmare. Take, for example, any attempt to link up a group of general practitioners across Canada. Entries in English and French, with different character encoding schemes, can exist that may have two sets of conventions, such difference adds further complication to the complex process of integrating with some legacy databases from different vendors.

Within a hospital, many applications may have been built for different purposes and each has its own database for a variety of information types. Integrating these applications for ease of exchange of information would ensure information about each patient can be shared while care is provided, and to assist better management. One example of integrating applications by connecting them to a shared database is shown in Figure 5.18. What matters most is consistency so that a single piece of information about a given patient from different healthcare providers can be accessed and updated simply by some kind of transaction management system.

5.3.3   Involving Different People

As in any system, a healthcare information system cannot be completed without people such as end users, maintenance support technicians, designers and engineers. They are different in how the system is perceived. The basic rule of thumb is retaining the user interface of all parts of a system in as original manner as possible. Alternation to the way users interact with the system should not be made unless it is absolutely necessary. This is particularly important with healthcare systems since there is no margin for error. Users should be able to continue using the system in the same way, before or after integration. Designers should therefore incorporate any new functions or features without changing the interactions to existing functionality. Furthermore, all interfaces should remain the same when integrating existing systems together. Successful system integration entails collaboration between installation engineers and both users and designers. Users need to be told of any temporary interruption that may be expected during the process in order to reschedule any tasks and to make alternative arrangements to cope with unexpected emergencies. Designers should ensure that anything that can possibly be done before shutting down an existing system will be done in advance so that any new module can be installed as quickly as possible to ensure minimal interruption.

Testing is a vital part of system integration as the process of testing enables any problems to be identified and rectified. Although we shall look at the details of system reliability and prognostics later in section 9.1, we'll conclude this section by taking a quick look at system testing for the sake of completion.

Integration testing is an extension of *unit testing*, where any modules to be integrated into an existing system are tested on their own before integration. Unit testing, prior to performing integration testing when carried out after putting everything together, ensures that any fault within a module can be detected and corrected prior to being put into an overall system that may otherwise cause serious damage to the system. Sometimes a module works well by itself but develops disparaging problems when integrated as part of a system. Exhaustive tests under all working conditions must be carefully carried out before and after system integration to ensure continuing reliability.

Compatibility is almost always an issue when adding new modules to an existing system. This is particularly niggling when integrating things from different manufacturers into an existing infrastructure. Standards conformation helps ensure compatibility and interoperability amongst devices made by different manufacturers. The abbreviation CII is often used in system integration with three different, but related meanings:

> *Common Integrated Infrastructure*: An integration model to integrate new and legacy applications within an enterprise for common cross-enterprise integration infrastructure (Helm, 1999). This is also applicable to healthcare enterprise systems.
>
> *Compatibility, Interoperability, and Integration*: A set of rules to ensure all these parameters are met as described above.
>
> *Configuration Identification Index*: A manifestation to correlate documentation to the proper set of configuration for individual systems.

The final piece of work is a user acceptance test, which verifies the system's performance and usability. This step is to ensure that the new system after integration matches all user requirements. User training may also be necessary to provide information on what has been added to the old system.

5.4   Evaluating IT Service and Solution Provider

Business opportunities are vast for IT companies seeking to enter the medicines and healthcare industry since technology can be applied to caring for everyone from head to toe. Different modes of partnership between healthcare providers and technology firms exist. Also, due to the large number of possibilities, there are a number of important issues to address. Readers should gain a broad understanding of what is involved so as to prepare themselves to optimize resources.

5.4.1   Outsourcing

Many IT related services are outsourced to third parties for both time and cost savings. It is very often true to say that money is best spent on letting the right people do the right job. People who are good at doing a specific task will accomplish work in the most efficient manner. This can be easily said but who can do it is another issue. Finding the right people may not be that straightforward after all. The IT industry is so huge that there are many people providing essentially the same type of services, and of course, some are good and some are bad. In cases of software development, outsourcing can even be offshore as the only thing needed is an Internet connection. Services such as diagnostic teleradiology, medical image processing, and electronic billing can be easily done by a company in a developing country where operating costs are much lower.

There are also many disadvantages and risks involved in outsourcing, the most obvious one being the risk of disclosing confidential information to the service provider. Also, competitors such as other clinics and institutions may happen to hire the same provider, it may not devote necessary attention to certain needs. Hidden costs may incur during the outsourcing process among other possible problems such as delay and misunderstanding.

Before commissioning someone for a service, we need to go through a checklist of performance measurement and determine what potential risks exist. Although different areas may have a very different set of items on the checklist, there are some general guidelines to follow. For example, when choosing a service provider for providing a wireless network, we need to look at parameters like bit error rate (BER), practical maximum data rate and other possible impacts on network performance when more users are connected. These are just a few examples of what should be checked against. These also include any after-sales support such as mean time for repair, guaranteed response time, any loan units available as substitutes while removed for repair, etc.

5.4.2   Coping with Emerging Technologies

Keeping up to date with what is happening in IT-related industries is extremely important since technology has a significant impact on optimizing operational efficiency and revenue cycle. With environmental friendliness in mind, modern products are designed to be more power efficient and the use of toxic substances is strictly limited. Emerging technologies are often associated with sustainability (Jablonski, 2009). This causes a number of other problems such as more restrictive design requirements and how a system is laid out in certain sites due to management or compliance issues. Some regulatory concerns may lead to additional costs or delay that a service provider should be fully aware of. The service provider should also advise what contingency plans may be necessary in the event of any change in government regulations or unforeseen impediments.

The ability of a service provider to keep updated with the market is extremely important. Imagine what happens if something has been planned for future use yet the service provider ends up having contracted to deliver something of obsolete technology. Advances of most IT products have such a fast pace that often makes them obsolete in a very short time. Below is a case study that looks at a portable glucose meter that was designed to be linked wirelessly to a home PC some two years ago. The development project was outsourced as the 'Vena' platform became available (Pordage, 2008), where compatibility of data exchange is supposed to be assured with appropriate procedures in place. Although there is nothing wrong with the development platform itself, the service provider did not realize that the IEEE 11073 standard was not yet finalized at that time. The glucose meter delivered failed to comply with the (IEEE 11073-10417-2009) standard as a consequence. The application profile standards were basically neglected during the product development stage. A device update is therefore necessary in order to communicate with PCs that are IEEE 11073 compliant. Such an update, of course, incurs service interruption and additional man power.

5.4.3   Reliability and Liability

Reliability can be judged in a number of ways, either quantitatively by some kind of metrics or something subjective such as word of mouth. Generally, a reputable company who has been in the industry for a long time can be relied on. In spite of this, we cannot always assume a good brand name always provides reliable service especially when supporting life-saving applications. Recall how many times our PC suddenly 'freezes' and we have to press Ctrl-Alt-Del to get it restarted. Not everything can afford the time for a system that suddenly stops responding for no reason. If we use something like this for resuscitation the chance is that the patient will not be revived by the time the unscheduled reboot process is completed. As reliability is such an important topic in medical technology, we shall take a closer look at this topic in the next section, as well as introducing prognostics for healthcare in section 9.1 where we shall look at some statistical analysis and modelling to determine the life expectancy of a medical device. For the remainder of this sub-section, we shall focus on how to determine if we can rely on a service provider in case we choose to contract out certain work to an external entity.

When we sign a contract to commission someone for a specific task to be completed on our behalf, we expect them to be reliable. This is exactly why we want to ensure that we assign the contract to the right person who we can depend on. Amongst a long list of service providers that we may find from different sources such as the Internet, perhaps through search engines, or the Yellow Pages, we need to shortlist a small number of potential suppliers through a screening process. Some guidelines should be drawn out according to a set of criteria. If we refer to the example in sub-section 5.4.2 above, we may want to lay down something like how long does it take to get the first prototype for trial, what standards (if any) will be adhered to, how can firmware updated be performed, whether the cost is within budget, etc. These are just a few of many items that need to serve as guidelines to choose the right supplier for us.

The first logical step would be to consolidate a long list of available suppliers so that we can hopefully end up with a dozen who appear qualified. An evaluation checklist should eliminate some of those who have no prior track record in carrying out similar work. From this initial process we should have names and capabilities identified as our shortlist after going through processes such as: product preference, service quality, operational coverage, and financial stability. In the business world, many companies would initiate an RFP (request for a proposal) when a number of potential suppliers are selected from the process. A qualification checklist should be prepared prior to sending out RFPs to the last few remaining candidates. Any specific requirements, the glucose metre must be wearable with battery life of at least 72 hours for example, must be investigated in more detail. Sometimes reference checks and enquiries to other institutions may be useful, although we should bear in mind that information acquired may be biased. A simple ranking process can position the potential suppliers in order of capability and suitability for our requirements.

All this tedious work is to ensure we can obtain reliable service. There is more to just finding the right supplier who can meet our requirements within budget. We need to not only make sure that our supplier is reliable, but also that the service that we will ultimately provide by using whatever we obtain from the supplier is reliable too.

Liability can be a very important issue, especially in the healthcare sector, where a patient may sue for damages that demand a huge sum of compensation should something go wrong. Defining who is responsible for what under certain situation is extremely important. This should normally be stated clearly on the service contract and checked by a legal representative. Legal liability, in the context of providing healthcare services, can be a very serious matter. Let us take a look at a simple illustration of a simple individual health insurance policy. This entails a long list of terms and conditions attached with maximum liability and what is excluded under what circumstances. Sometimes a liability waiver is required before providing certain healthcare services. This is to limit any risk of being held responsible in the event of an emergency. A simple waiver form like that shown in Figure 5.19 would ensure written authorization is given when attending to an emergency.

It is worth noting that liability issues exist between the service provider and any supporting entities, such as outsourced contractors, equipment manufacturers, solution architects, as well as with patients. Ultimately, we do not want to end up in a situation where we are held responsible for any issue caused by our service providers. To prevent putting ourselves into any undesirable situation, quality assurance would be what we want to take good care of, and we shall take a close look at the details in the next section.

5.5   Quality Measurement

Quality is perhaps the most important attribute in providing trustworthy healthcare services. Remember, we seek to provide eminent healthcare with wireless telemedicine technology. We shall look at what can possibly go wrong with wireless communications in order to maintain quality of service (QoS) by investigating major factors that can impair wireless communications.

The term 'link outage' refers to the situation where a wireless channel is cut out. Communications and information theory is often accompanied by a statistical model that describes the probability of the successful reception of some kind of information, as it goes through the model of Figure 2.1, by a receiver. The information is sent out from a transmitter across a wireless channel. Main factors that can cause problems to the propagating signal (the signal that carries the information across from the transmitter to the receiver, through the channel) include (Fong, 2003b):

- Attenuation: weakening of signal strength over distance travelled
- Depolarization: reduction of separation between two signal paths of different polarizations resulting from phase retardance
- Interference: disruption of signal caused by other sources
- Noise: unwanted additive energy that is inserted into the signal
- Scattering: radiation towards different directions after hitting an object

MEDICAL/LIABILITY WAIVER FORM
This form must be signed by the patient before receiving medical service.
This form will be securely stored for up to six months.

Last Name: _____ First: _____ DOB: _____

Parents/Guardian (if below 18): _____

Address: _____

_____

Phone: _____

Email: _____

Medical Information and Release

In the event of an emergency, who should be contacted?

Name: _____ Relationship: _____

Phone (Home/Office): _____ Mobile: _____

GP: _____ Phone: _____

Which hospital would you prefer you and/or your child be taken to in case of an emergency?

_____

Medication currently taking and/or known allergies:

_____

_____

Other relevant medical information:

_____

**In the event of an emergency, if neither parent nor emergency person(s) can be contacted or if
there is no time to make such contact, the following signature authorizes such emergency
medical and surgical treatment to be provided, including transportation to the nearest facility,
as may be deemed necessary.**

Signature _____

Date _____

Full Name: _____

Figure 5.19   A sample medical liability waiver form

There are many other signal degradation factors, too. To illustrate the complexity of signal
degradation, we take a look at interference. Interference that can affect the reliability of a
communication system may include (Stavroulakis, 2003):

- Co-channel interference: also known as 'crosstalk', effects of encountering signals from an
  adjacent channel of similar frequency.
- Electromagnetic interference: also known as 'radio frequency interference' (RFI). Interruption due to signals from other sources, This is a technique intentionally used for *radio
  jamming* so that one can disrupt a wireless link by emitting another signal of similar frequency. In severe cases, solar radiation may also cause RFI in rare cases.

- Intersymbol interference: unwanted interaction between adjacent symbols (of data), usually caused by multipath, the net effect is quite similar to noise that is caused by the same signal of slightly different time.

As we can see, interference alone has several different kinds. To measure the quality of a given wireless link, we have different parameters, these include:

Bit Error Rate (BER): measures the number of error bits that occur within a block of bit stream. For example, $BER = 10^{-6}$ means statistically we can expect one corrupted bit per one million bits transmitted. This figure is normally considered as acceptable for general consumer electronics applications. However, telemedicine requires better quality than this (Schimizu, 1999). To improve BER performance of a given wireless link, reduction in data rate or allocation of adequate *link margin* can be considered. The link margin refers to the extra power necessary to combat signal loss due to different degradation factors. For example, a certain link margin is necessary to allow for attenuation due to rainfall. Since BER is a measure of data bit error, it is a performance measure against the E_b/N_0 (the energy per bit to noise power spectral density ratio) value for a given channel. E_b/N_0 can be viewed as the digital equivalent of the Signal-to-Noise Ratio (SNR) in analogue communication systems, or more appropriately, the SNR per bit. E_b/N_0 increases as BER improves (i.e. the BER value decreases, e.g. from 10^{-6} improves to 10^{-9})

BER is usually measured by a BERT (bit error rate tester), often in the form of a software package.

Signal-to-interference ratio (SIR): measures the ratio of signal power to that of the interference power in the channel to check the received signal quality at a receiver. It is sometimes called the 'carrier-to-interference ratio'. SIR is similar to the SNR of a propagating signal before it is processed by the receiver. In this respect, the main difference between the interference 'I' and noise 'N' is that the former originates from an interfering transmitter source which is controllable through network resource management; whereas the latter comes from a combination of many manmade and natural sources. SIR should normally be at least 18–20 dB to ensure quality reception. SIR is usually improved by appropriate filtering algorithm for the receiver.

Statistical modelling is usually used to measure the signal power in order to measure SIR (Wang, 2001). This is essentially a process of developing an algorithm to analyze the signal power at the receiver before demodulation in relation to that of all interfering signals.

Carrier-to-noise-and-interference ratio (C/(N+I) or CNIR): is a measure of the amalgamate effect of both noise and interference in the context of CIR and SNR.

Co-channel interference (CCIR): nearby channels operating at the same frequency that cause interference between each other. Increasing the SNR not only will not improve the impact on interference, it can make the situation worse. Reduction of co-channel interference can be done by increasing distance between co-channels (Chen, 1997). In the USA, FCC regulates the 'out of band' noise for radio transmitters in order to suppress sidebands that cause interference. To combat this problem, input filters with sharp cut-off are usually deployed at the receiver.

Link outage: a statistical measurement of how much time within a year that a wireless link is cut off. This is enumerated in minutes and seconds. For example, a system with 99.99% *availability* would have a maximum link outage or down time of 52 minutes per year. This is calculated by a simple equation as follows, given that there are 31 536 000 seconds in one year and t is the maximum permissible link outage time. In this particular example with 99.99% availability, Equation 5.10 gives us 3 153.6 seconds so we simply divide this by 60 to convert our annual permissible outage to just over 52 minutes.

$$t = 31536000.(1 - availability\%) \qquad (5.10)$$

We have discussed a number of major measurement parameters to quantify the quality of a wireless system. Ultimately, quality measurement ensures a wireless telemedicine system is capable of supporting its services. In most cases, practical systems will perform somewhat worse than what is theoretically computed due to many uncontrollable factors.

References

Chen, J. Y. and Siu, Y. T. (1997), On co-channel interference measurements, *IEEE International Symposium on Personal Indoor and Mobile Radio Communications, Helsinki,* Conference Proceedings pp. 292–296.

Fong, B., Rapajic, P. B., Fong, A. C. M. and Hong, G. Y. (2003a), Polarization of received signals for wideband wireless communications in a heavy rainfall region, *IEEE Communications Letters,* 7(1):14–15.

Fong, B., Rapajic, P. B., Fong, A. C. M. and Hong, G. Y. (2003b), Factors causing uncertainties in outdoor wireless wearable communication, *IEEE Pervasive Computing,* 2(2):16–19.

Fong, B., Ansari, N., Fong, A. C. M., Hong, G. Y., and Rapajic, P. B. (2005), On the scalability of fixed broadband wireless access networks, *IEEE Communications Magazine,* 42(9):512–518.

Gibson, P. G. (2002), Outpatient monitoring of asthma, *Current Opinion in Allergy and Clinical Immunology,* 2(3):161–166.

Helm, R. (2000), Extending EAI beyond the enterprise, *Journal of Enterprise Application Integration,* http://www.eaijournal.com/article.asp?articleID=266

Hummel, S. (2007), *Cisco Wireless Network Site Survey,* BookSurge Publishing USA, ISBN: 1419667491.

IEEE 11073-10417-2009 (2009), Health informatics-personal health device communication part 10417: device specialization- glucose meter, *IEEE Standards May 8 2009,* ISBN: 978-0-7381-5894-5, http://ieeexplore.ieee.org/servlet/opac?punumber=4913383

ITU Recommendation X.200 (1994), Information technology - Open Systems Interconnection - Basic Reference Model: The basic model, Art. E 5139.

Jablonski, C. (2009), Futurist pinpoints world's top ten long-term challenges, *ZDNet blogs June 2009*: http://blogs.zdnet.com/emergingtech/?p=1607

Krakowiak, S. (2009), Middleware Architecture with Patterns and Frameworks, http://sardes.inrialpes.fr/~krakowia/MW-Book/

NIST (2006), General Purpose Link Budget Calculator, *National Institute of Standards and Technology,* http://www.itl.nist.gov/div892/wctg/manet/prd_linkbudgetcalc.html

Pordage, P. (2008), Groundbreaking platform allows medical devices to communicate wirelessly, *Cambridge Consultants White Paper* (UK 25 March, 2008).

Rimassa, G. (2002), Wired-wireless integration: a middleware perspective, *Internet Computing,* 6(5):96.

Schimizu, K. (1999), Telemedicine by mobile communication, *IEEE Engineering in Medicine and Biology Magazine,* 18(4):32–44.

Sheikh, K., Gesbert, D., Gore, D., and Paulraj, A. (1999), Smart antennas for broadband wireless access networks, *IEEE Communications Magazine*, 37(11):100–105.

Shmueli, G., Minka, T., Kadane, J. B., Borle, S., and Boatwright, P. B. (2005), A useful distribution for fitting discrete data: revival of the Conway-Maxwell-Poisson distribution, *Journal of the Royal Statistical Society: Series C (Applied Statistics)* 54(1):127–142.

Spahni, S., Scherrer, J. R., Sanquet, D., and Sottile, P. A. (1999), Towards specialised middleware for healthcare information systems, *International Journal of Medical Informatics*, 53(2–3):193–201.

Stavroulakis, P. (2003), *Interference, Analysis and Reduction for Wireless Systems*, Artech House Mobile Communications Series, ISBN: 1-58053-316-7.

Wang, C. W. and Wang, L. C. (2001), Signal to interference ratio measurement techniques for CDMA cellular systems in a frequency-selective multipath fading channel, IEEE Third Workshop on Signal Processing Advances in Wireless Communications, Conference Proceedings ISBN: 0-7803-6720-0, pp. 34-37.

Zou, Y. and Chakrabarty, K. (2005), A distributed coverage- and connectivity-centric technique for selecting active nodes in wireless sensor networks, *IEEE Transactions on Computers*, 54(8):978–991.

6

Technologies for Safeguarding Medical Data and Privacy

Information has been a precious asset to society since the existence of mankind. Thousands of years ago, ancient people started sharing information about where to find food and shelter. As society became more complex, some information was shared as knowledge, while some was kept with strict confidence for security reasons. For example, books exist for the purpose of sharing existing knowledge and to enlighten new ideas built upon known facts; bank vaults are designed for locking up private effects so that whatever stored inside is only accessible by authorized persons. In the past, most medical information is stored in physical formats such as cards and logbooks. As a result of rapid expansion in the amount of information being collected and created, there are more incentives to use computer-based data storage media for safe-keeping of medical information. The topic 'Information Security and Privacy' is simply the course of protecting information availability, data integrity, and confidentiality, so that it will only be accessible to authorized personnel, data cannot be tampered with, and it will not be leaked out.

We have talked about the importance of data security and privacy from time to time throughout the earlier chapters. Its significance needs no further discussion as it should be well understood by now. There are two main rationales, either keeping information related to an individual in strict confidence, for example, medical history of a patient; or collecting anonymous data for statistical analysis, for example, by conducting a healthcare survey; it is vitally important to ensure that any data collected cannot be used to identify a person or where it comes from.

Many countries already have legislations governing the privacy of individually identifiable information and the confidentiality of electronic patient records so that information access is strictly limited to authorized personnel with the consent of the patient, for example, the Health Insurance Portability and Accountability Act (HIPAA) in the USA. In contrary, regulations governing patient privacy in the UK are moving towards the use of such information without patient's consent (Thomton, 2009). Such initiative causes even greater concern to security and privacy of health information. In this chapter, we shall look at security and privacy in two areas, namely safeguarding patients' medical history and the use of biometric features for identification. The former is important for the interest of the general public, whereas the latter

Telemedicine Technologies: Information Technologies in Medicine and Telehealth Bernard Fong, A.C.M. Fong, and C.K. Li
© 2011 John Wiley & Sons, Ltd

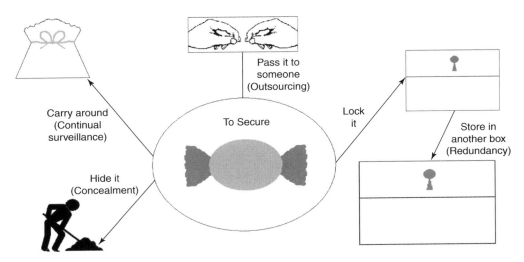

Figure 6.1   A simple safe-keeping plan

is technology widely used in personal identification such that an individual can be uniquely identified.

6.1   Information Security Overview

Information security involves compromise between security and usability. To illustrate this, we look at an example where a little girl called *Melody* wants to safeguard a candy from being taken by others. Melody's idea is to make it as safe as possible so that she has drawn up a simple plan in Figure 6.1. Melody realizes that if she places the candy in a treasure box, and this box is then concealed inside a bigger box, there is a smaller chance that the candy will be found by others. Alternatively, she can dig a hole to hide the candy underneath the ground. She realizes that the deeper she digs the smaller the chance of it being taken by others. So, Melody has hidden the candy in a safe place. However, when she wants to eat the candy she finds that it becomes more troublesome and time-consuming to retrieve the candy. From this scenario, Melody learns that having more security in place will make it more difficult for someone to get access. It will also take longer to be retrieved. Information (or data) security works in exactly the same way.

Now, Melody decides to put her candy into a little pouch and she passes the pouch to her brother Vale. Vale then slips the pouch into his bag and this is passed on to their mother. While mother drives the children to the mall, Melody decides to eat her candy. She asks Vale for the pouch, then as Vale gave his bag to mum he asks mum for the bag. As mum is concentrating on the road she asks Vale to open the armrest storage compartment for the bag. So he does and puts his hand into the bag, after a few seconds he holds Melody's pouch and passes it to her. Melody suddenly shouts 'where is my candy?' What has happened? Well, to help Melody we investigate every possibilities of why the zipper is found opened. Did Vale open it or did Melody forget to zip her pouch? Was the bag packed and handled by Vale properly? Or, has mum done anything to it? Here, we can see in this simple example that 'Security is Everyone's

Responsibility'. No one in the system can deny a part when ensuring security. In this example, although only Melody deals directly with security, everyone can be held responsible for the loss.

Owing to the famous quote 'a chain is only as strong as its weakest link', security is weak if there is one single point that exhibits any form of weakness no matter how strong a security system is. Therefore, security relies on everyone to safeguard everywhere.

6.1.1   What are the Risks?

IT applications, including those which support medical and healthcare services, often need to meet the conflicting requirements of users. Problems such as secured access to data and applications may arise when users of different roles attempt to share something, such as in the example of police requiring information about a patient's medical record for criminal investigation. How to share the information, whether direct access should be allowed and what to share are simple questions that need to be asked. This kind of situation, when different users have different requirements due to different perspectives, may cause security issues that make a system more vulnerable to attacks. Security risk is related to both the likelihood of a security breach of any form and its impact. There are many types of risks, some more serious than others. Among these dangers are viruses that can erase everything in the system, breaking into your system and tampering with the stored data, someone impersonating you or using your computer to assault others. Unfortunately, it is an unrealistic expectation to guarantee with absolute certainty that these will not happen despite all the best provisions. We can only do whatever necessary to minimize the risk and any consequential impact. Although we cannot totally eliminate the risks, there are ways to control or manage them with appropriate policies, procedures and practices; involving management legal, technical, and administrative aspects (Rindfleisch, 1997). Before we go further, here is a brief discussion on some common terms:

> Hacking: Activities that explore weaknesses in software and computer systems. Some may have benign intents motivated by curiosity while others may have criminal engagement of stealing or altering data.

> Malware: also known as *Malicious* code, is a small piece of software written for attacking a computer. These are things like viruses, worms, and Trojans.

> Phisher: Spear phishing is an e-mail spoofing fraud attempt that targets a specific group of people or an organization, often to gain unauthorized access to secured information.

> Spam: Detestable advertising materials that flood the Internet and waste network resources such as bandwidth and mailbox storage with junk, usually sent with malware.

Having talked about these manmade events, we cannot overlook risks caused by natural phenomena such as storms, flooding, fire, and earthquake. Since the early age of computers people have been very conscious about data backup. Backup is the process of making exact copies of the data in another storage medium so that the data can be retrieved in the event of a loss or failure of the original copy. In the past, bulky backup tapes such as those shown in Figure 6.2 were used for archival of a periodic basis. There were a few major problems with

Figure 6.2   Obsolete backup tapes used for decades in the past

these tapes in addition to the slow data retrieval speed, as shown in Figure 6.2 many tapes are stacked in a cabinet so they occupy heaps of space. Another problem about storage is that magnetic tapes are prone to humidity and mould therefore these tapes were usually stored in a controlled environment where the temperature and humidity remain more or less unchanged. Before networking became popular off-site redundant storage was a logistical nightmare as frequent update of each backup copy made storage in a different site extremely impractical. Imagine what you need to do if you have to distribute tapes to a few locations on a daily basis. Storage in more than one location is extremely important for prevention of fire and flooding. In the event that a fire breaks out you still have another copy somewhere.

In a networked system, frequent data backup in various locations is very easy. As shown in Figure 6.3, data is simply sent to *mirror site* backup facilitates via the network with appropriate synchronization. As its name suggests, a *mirror* is an exact copy of the data in computer terms. So, a *mirror site* is simply an exact duplicate of another site. All mirror sites can be synchronized to be automatically updated once the original data changes. In case the duplicate server breaks down, the mirror server can operate in its place so that data retrieval will not be disrupted.

Data vulnerability can be related to main areas within the entire communication system. The weakest link can be anywhere in the system. Repeated cases of careless hospital staff members losing their USB thumb drives containing patients' information have been reported in (ComputerWorld, 2009). Such an irresponsible act can make even the strongest security system useless.

Security also depends on network configuration. In a peer-to-peer network, there is normally a trust between servers so that a user who has access to one server will automatically be granted access to another. An intruder can therefore move freely throughout the network once access to one server is gained.

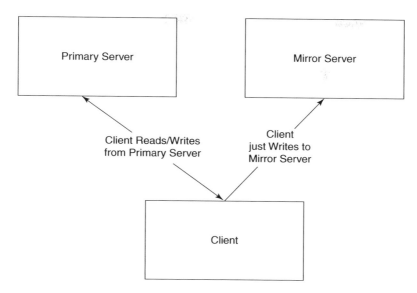

Figure 6.3   Back up with a mirror site

6.1.2   Computer Virus

The threat of cyber-terrorism has been expanding rapidly since the Internet era over the past one to two decades. Someone opening an email attachment risks spreading a virus throughout the entire hospital network and beyond. Some viruses can unleash themselves without even opening the host file itself. Just like viral infection in the human body, a computer virus can sneak into the system, spread to other computers by copying itself. Since computer virus are willfully created by people, it can mutate, be of destructive, malicious, or bothersome in nature.

Similar to (biological) viruses, there are other software codes that people create for depraved reasons. These include:

Worm: a self-replicating program that spreads itself across the network. Its main difference from a virus is that it is self contained, whereas virus attaches itself to another program or file, such as a script or an image. Also, most viruses are written to attack computers, whereas worms are designed to attack networks.

Spyware: software written to observe the interaction between the computer and its user and sends such information to a third party via the network. This can be risky as it can also 'steal' data, including confidential information stored.

Trojan: malware that disguises itself in perceivably harmless application software that includes code to allow access to a computer and its stored data. This can cause serious consequences such as stealing information and seizing control of the computer.

6.1.3   Security Devices

Many devices are available for making a network more secure. These may include purpose-built devices or software installed on a computer. In a communication network, there are many places where we can install something to safeguard security, and this something can perform different types such as identifying an individual user, granting access to part or all of a system, logging activities during a session, and to filter incoming and outgoing data based on types, origin or destination, inclusion of certain keywords, etc. To understand more about the features of security devices, we shall look at some commonly used security devices:

> Firewall: a rule-based device that filters certain types of data from entering a network. A firewall can be implemented either as a hardware box plugged into the network, or as software installed on a computer; it can also be a combination of both.

> Front End Processor: a host computer that manages the lines and routing of data in the network, it can also authenticates a user when attempts to login are made from a remote location.

> Proxy Server: a type of firewall operation that specifically filters out anything that either enters or leaves a network. It essentially hides the actual network addresses so that any attack will be made more difficult without knowledge of information about the network.

So, all these devices have one thing in common, either allowing or denying access of users or data from passing through a network. This may sound simple enough by applying a number of rules, but in the real world, implementation of security plans are far more complex than just setting up some security devices as damage to a network can be carried out from a large number of locations, as we shall discuss below.

6.1.4   Security Management

Having looked at some basics of information security, we should have a broad understanding that security is about management of:

> Integrity: data remain in the form it should, without being tampered with in any way.

> Privacy: patient information is not released to unauthorized entities.

> Confidentiality: safeguard personal and corporate information, assets such as patient records, development plans and work schedules are all kept with strict confidence.

> Availability: system is kept available at all times, without being affected by sabotage or breakdown.

To facilitate all these, security needs to be addressed in the following areas:

Computer system security: hardware and software, access to computers and data stored in them. Protection against virus infection is also an important issue to address.

Physical security: areas where equipment is installed should be monitored. This applies to both restricted and open access areas. For example, doctors should not leave their computers unattended so that someone can sneak in with a USB drive to copy data from it. Protection of portable equipment is an increasingly important topic, laptop computers are stolen by organized criminals for the data stored inside (Sileo, 2005).

Operational security: operating conditions and usage logging, conditions include ensuring clean power supply, such as using an uninterruptable power supply (UPS). A UPS has a mechanism to ensure that power is not interrupted and still available for a short time even in the event of a power failure, so that users are given sufficient time to save all data before performing a proper shutdown. In addition to serving as a short-term power supply, a UPS also filters out any glitches in the power source so that the output power remains clean. A typical UPS has an external battery and power filtering mechanism to ensure any power surge is removed. Figure 6.4 shows an example with two boxes, a power outlet and an outboard battery. UPS for computer systems usually have accompanied control software so that features such as automatic file saving followed by shutdown can be configured. Also, a fairly accurate estimate of remaining battery life can be displayed on the screen when backup power is activated and users will be alerted to a mains power cutoff.

Communications security: protection of network and communication equipment including computers, routers, and PDAs; network access ports as well as those possible attack points listed in sub-section 8.1.4 should be monitored.

Privacy and confidentiality leads to the issue with information classification, data can be classified into different categories according to risks, data value, or any specific criteria. Information may have different value or use; and is subject to different risk levels. We should therefore implement different protection procedures for different types of information. For example, the consequence of leaking patient records will be far greater than that of having information about drug usage stolen.

The use of proper preventive measures reduces the risk of security attacks. Information security management involves a combination of prevention, detection and re-action processes. Contingency plan is a vital piece of document that details how to respond to a threat and how to combat a problem when it arises. Proper security management would ensure risks from all sources can be minimized even though they cannot be eliminated.

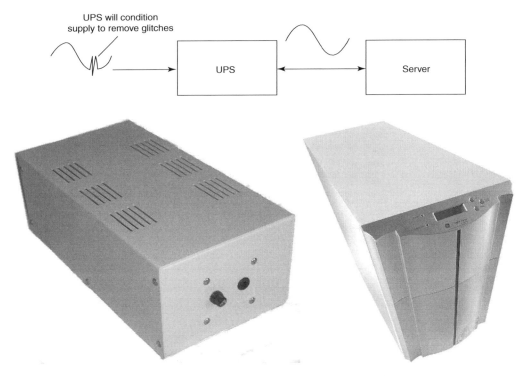

Figure 6.4   Uninterrupted power supply with external battery

6.2   Cryptography

Cryptography is the process of converting meaningful data into a scrambled code for transmission across any communication channel that can be 'deciphered' or converted back into the original data. The primary function of cryptography is to hide the original information so that it appears to be meaningless while on transit. It is such a vast topic that Schneier (2007) presents a set of three volumes totaling 1 664 pages exclusively on cryptography. Cryptography involves applying some kind of algorithms to convert the data before transmission, and when the 'encrypted' data reaches the receivers, it needs to be 'decrypted' back to its original data for interpretation. The process of encryption followed by transmission and subsequent decryption is illustrated in Figure 6.5, where a *key* is generated (a number code) and distributed to both the transmitter for encrypting the original message (expressed in plain text) and the receiver for decrypting the ciphertext to extract the original message. Although cryptography may not be able to achieve absolutely 100% security, it does serve as an essential part of a secure communication system due to its effectiveness and capabilities. It is very widely used in virtually all aspects of communications, including but not limited to patient records, medical images, supply order processing, e-prescription, transactions processing, etc.

Referring back to Figure 6.5, Melody wants to send a secret to Vale, without letting mum know. They first agree on a key pair $k = (c,d)$ so that both Melody and Vale keep a copy of the same key. When Melody sends her secret message m to Vale, she uses the key k to generate

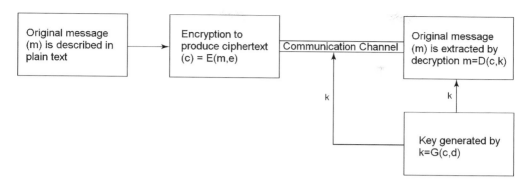

Figure 6.5   Cryptography

the ciphertext $c = E(m,k)$ and send c to Vale. She knows that the message m by itself can be read by mum, and if mum picks up c it will make no sense to her since she does not have the key to 'decode' the message from c. When Vale picks up c, he uses the inverse method that Melody used to encrypt the message with the key, so that $m = D(c,d)$ to extract the original message m from the received encrypted message c. Melody trusts this method because she knows mum does not have the key or any knowledge about the method she used to generate c.

So, here is the basic principle; but why does Melody need a key? On hindsight it may appear as if Melody can just choose any encryption method and Vale applies its corresponding decryption method. Here, the main purpose of the key is that even if mum finds out the method they use, they can still use it without redesigning the method very simply by using a new key. It is therefore only a matter of keeping the key changed every now and then.

Loosely speaking, there are two approaches of encrypting data, either symmetrical or asymmetrical. The former uses the same key for encryption and decryption, whereas the latter uses one key for encryption and another different key for decryption. In this section, we shall look into the details of both approaches. To illustrate how these algorithms work, we shall bring in our two children, Melody and Vale, to explain to us the underlying mechanisms.

6.2.1   Certificate

A *digital certificate* is an electronic document for the identification of a user or a server. Like any form of personal IDs, a digital certificate serves as a proof of personal identify. It is issued by a certificate authority (CA), which is an entity that validates identities and issues certificates. The CA is just like any government agency that issues personal IDs, where certain checks are performed to authenticate a person's identify before issuing a valid ID for that person. Methods used to validate an identity can be different depending on individual CA's applicable policies. This is similar to different policies imposed by UK's Driver and Vehicle Licensing Agency (DVLA) for driver licensing and the Identity and Passport Service for ID cards and passports.

Client authentication is the process of identifying a client by a server so that the identification of a user on the client can be checked; whereas *server authentication* is the opposite process where identification of a server is verified by a client so that a user can be assured that the server

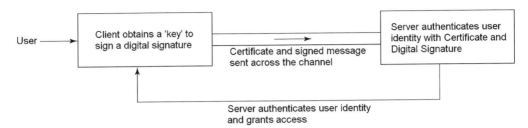

Figure 6.6   Certificate-based authentication

is indeed one that the user want to access, such as in the case of ensuring that a bank's website is legitimate but not one that is forged by criminals who attempt to steal login information.

Certificate-based authentication is widely used in the Internet where the client signs a *digital signature* (see section 6.2.4) and attaches the certificate to the signed data to be sent across the network. The server validates the signature and the certificate upon receipt. The entire process of certificate-based authentication is shown in Figure 6.6. Often used in software distribution and electronic patient records, digital signature is a code that proves the authenticity of a message. Digital signature uses *asymmetric cryptography* for messages sent across an insecure network to ensure the true identity of a sender.

6.2.2   Symmetric Cryptography

Also known as *private key* or *secret key* encryption because the key used is never made available to any parties other than the sender or receiver. As shown in Figure 6.7, its operation is fairly simple as we look at an example here. Suppose Vale sends a message to Melody using private key encryption, the process is as follows:

1. Melody creates a key and sends a copy of this key to Vale
2. Vale uses this key to encrypt his message
3. The encrypted message is sent to Melody via the network
4. Melody gets the encrypted message, she decrypts it with the key

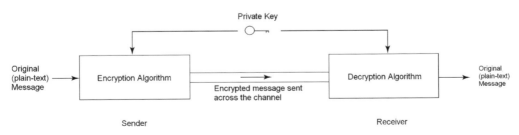

Figure 6.7   Private key encryption

This mechanism is fast and simple to implement as Melody only needs to generate one key and send it to Vale for encrypting the message. The first obvious problem with this mechanism is that Melody has no way of checking the integrity and authentication here. So, from the received message alone she cannot find out if the message has been corrupted or if it was indeed sent from Vale. Also, Melody has to advise Vale in advance which key to use. They have to have the identical keys on both sides for this to work. To overcome these fundamental problems, most modern key systems use the *asymmetric* approach.

6.2.3   *Asymmetric Cryptography*

Also known as *public key* or *shared key* encryption because a key is generated and placed in the public domain so that potentially anyone can get this key. The term *public key* refers to encryption that uses a key that is published so that it is basically available to everyone. Since the public key (for encryption) is used to 'generate' a *private key* for decryption, everyone therefore has a pair of different keys. Anyone can publish a public key so that whoever wants to send a secret message to the person who publishes the public key can do so.

To look at how public key encryption works, we look at the mechanism behind when Vale sends a message to Melody:

1. Melody generates a public key and places it on the table (everyone can access this key because the table is insecure)
2. Vale grabs the public key from the table
3. Vale uses this key to encrypt the message.
4. The encrypted message is sent to Melody via the network
5. Melody gets the encrypted message, she decrypts it with her private key (the private key has never been made available to anyone, including Vale)

This process is summarized in Figure 6.8, where we can see the private key is securely stored in the receiver. Melody has two keys, a private key that she keeps and a public key that she places in an insecure place that anyone can access. The public key is only used in encrypting the original message. The encrypted message is sent out, without any key attached, to the receiver and the receiver then uses the private key for decryption. The process of key generation is summarized in Figure 6.9.

Figure 6.8   Public key encryption

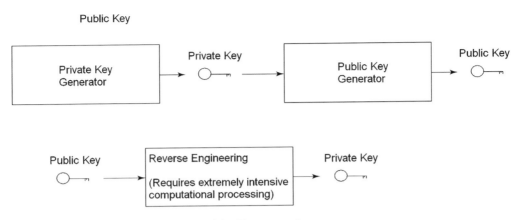

Figure 6.9   Key encryption process

The main advantage of this key system is to avoid the risk of having a shared secret key being stolen when it is passed over a communication channel. So, a public key is originally designed to eliminate the need for exchanging the key over a communication network. The public key concept was originally developed by British mathematician Clifford Cocks in 1973; the original work does not have a reference as it was classified as a British government secret. The public key system was later formally defined by Ron Rivest, Adi Shamir, and Leonard Adleman in 1978 (Rivest, 1978), commonly known as the RSA algorithm, named after the three developers' surnames.

In a public key system, everyone who is connected to the network can access the public key, so that this key can be used to encrypt a message for sending the encrypted message to the person who publishes the key. Only the receiver can decrypt the message with a private key that is not available to the public. The system's main feature is that the secret key does not need to be sent through the network, thereby eliminating the risk of being stolen while in transit. This system eliminates the need for agreeing on which key to use, as it will always be the one that has been made publicly available. The message will remain secure as long as the receiver keeps the private key secret. Public-key cryptography uses certificates to uniquely identify a person therefore authenticity can be guaranteed. However, the system is prone to plain text attacks such that such encryption can be decoded by hackers because the private key (for decryption) can be generated by using the public key that everyone can obtain. While such threats can be minimized by proper design and implementation of the cryptographic process, it is in fact possible to generate the private key from the public key using some algorithm with the necessary intensive computational power. So, public key cryptography by itself is not 100% foolproof. The principles behind breaking the public key system are not within the scope of this text, an overview of attacks on the RSA system is given in Wong (2005).

6.2.4   Digital Signature

Public key encryption uses digital signatures for data integrity assurance. The digital signature, in the form of data codes, is attached to a message. It can be used to check whether the message

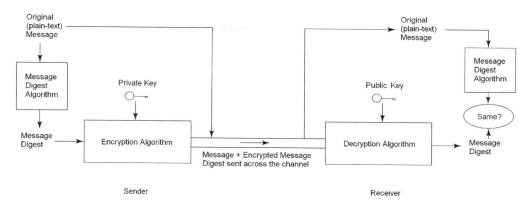

Figure 6.10   Digital signature

has been tampered with during transmission. The sender uses a unique signature so that the message is encrypted with a *message digest* algorithm. The set of codes that forms a digital signature is computed based on both the message and the sender's private key.

Like a hand-written signature, digital signature relies on the slim statistical probability that two identical signatures created by different entities will never exist. When the public-key system is used to generate a digital signature, the sender encrypts a *digital fingerprint* based on the message along with the private key. The signature can be verified with the pubic key by anyone. To see how this works, Vale wants to send a signed message to Melody. The process illustrated in Figure 6.10 is as follows:

1. Vale generates a message digest by using a *message digest algorithm* applied on the message
2. Vale encrypts the message digest with his own private key to generate a digital signature
3. Vale sends the message with encrypted message digest attached
4. Melody authenticates the signature by applying the same message digest algorithm
5. Melody decrypts the message digest using Vale's public key to compare with (4)
6. Digital signature verified if the results of (4) and (5) are identical
7. Authentication fails if (4) and (5) do not match, this tells Melody that the message is either sent by someone who impersonates Vale or the message has been tampered with

An obvious advantage of creating a digital signature by encrypting only the message digest but not the entire message itself is computational speed, as the message digest is much shorter than the message. The major problem is the possibility of *collision*, which occurs when the sender signs other message with the same message digest, the situation when there is more than one message having the same message digest is known as collision. Message digest algorithms should be designed to avoid collision.

6.3   Safeguarding Patient Medical History

Recognized by the UK parliament, electronic patient record (EPR) systems can benefit both patients and practitioners by improving clinical communication efficiency, reduce errors, and

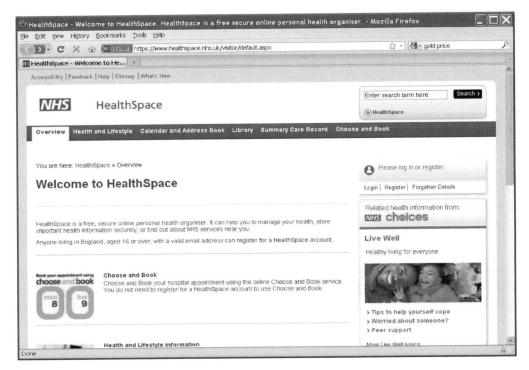

Figure 6.11   Screen shot of the NHS HealthSpace website

assist in diagnosis and treatment (Barron, 2007). The report describes National Care Record Service (NCRS) that creates two EPR systems, the national Summary Care Record (SCR) and local Detailed Care Records (DCRs). As their names suggest, the SCR contains general information for the entire UK, whereas DCR contains all-inclusive clinical information in a local context. Almost a year after the launch of the NCRS, (Greenhalgh, 2008) presented a study on patients' attitudes to the SCR, which found that the general public was unclear about the policies on shared records while most surveyed viewed the system as a positive development. The SCR is now available for UK residents via the *HealthSpace* website. Anyone living in England of age 16 or over can register for a Basic Account, which lets users book a hospital appointment. Registration for an Advanced Account would require participation of NHS Care Records Service by a user's local NHS. Only the Advanced service grants users access to their SCR.

The NHS HealthSpace website provides a range of services. Referring to Figure 6.11 which shows its homepage, users can scroll down for the following services in addition to account registration:

- Booking a hospital appointment even without an account
- Health and lifestyle information on various parameters such as blood pressure, cholesterol levels, and medications
- Keeping appointments and location of clinics, pharmacy, and NHS offices
- Access to SCR by logging in with an Advanced Account

In the above case study, we have seen the UK health authority has already implemented EPR even though there may still be room for improvement on feature enhancements and time for information update. Currently, the system still requires SCR to be manually set up by an Advanced Account holder's GP.

6.3.1 National Electronic Patient Record

The NHS HealthSpace provides residents in the UK with a tool for accessing their own health information online, and individual's details are also accessible through their healthcare service providers. Important information for emergency treatment, such as drug allergy and ongoing therapy can be retrieved when needed. Owing to the useful features of EPR, the Social Insurance Institution of Finland is also introducing a system similar to HealthSpace. Finland's main feature is the inclusion of a telemedicine system for medical image archive. Entries by all medical professionals across Finland will append information onto a patient's archive. It also supports automated delivery of prescriptions which assists with the prescribing and dispensing processes.

National EPR systems do have their drawbacks, problems with participation has been reported in The Netherlands (Weitzman, 2009). Comments posted on *researchblogging.org* suggests that 31% of Dutch doctors are reluctant to subscribe to the national EPR with a further 25% considering doing so. Given the low ratings, what is wrong with the Dutch system? The first fundamental problem with their system is a virtual EPR, meaning that medical data will remain physically where it originates but not in a national server. This therefore requires all participating doctors to have their services accessible online at all times. Linking up servers across all clinics would mean there will be a cost involved for the software and hardware for network connection, which the Dutch authority does not provide full subsidy on. In addition to this problem, the system is initially implemented as an amalgamation of two separate units, an electronic medication record and a deputy GP record. The former stores information about each patient's visit including prescription details, and the latter provides after-hours access of patient data. The arrangement is prone to unauthorized access to individual patient's data since anyone who can access a clinic's computer can access all patient records. In theory, patients' privacy is respected by means of contacting a supervisory agency in the event of a suspected infringement of privacy. However, such reporting method requires patients to initiate an enquiry after a patient suspects their own data has been accessed without consent. This would be virtually impossible in practice since no mechanism exists that alerts a patient of any access to data. More information about the Dutch national EPR can be found in Tange (2008).

6.3.2 Personal Controlled Health Record

A Personal Controlled Health Record (PCHR) is a form that a patient can use to control their access rights and contents. User control is accomplished by subscription and appropriate access control mechanism such as by using password access. The system enables patients to own and manage a complete and secure electronic copy of their medical records. Patients can choose to connect their record to entities such as clinics and pharmacies at will, this can improve the management and analysis of their own medical data.

A similar approach is the Personal Health Record (PHR), accessible through the Internet for a compilation of EPR containing a wide range of medical information and history as well as other personal information such as age, address, etc. The two well-known systems are *Google Health* and *Microsoft HealthVault*.

6.3.3 Patients' Concerns

The sheer size of the national EPR database may cause concern to many as to whether their data is secured stored. Looking at Finland's example again, its population of 5.2 million is expected to occupy as many as 500 petabytes (each petabyte is equivalent to 1 024 TB). We are roughly talking about half a million 'huge' (as of end-2009) 1 TB hard disks to store the Finnish EPR. Roughly speaking, this is about 100 GB of storage for each patient. Security is always the biggest concern. In the Finland case, access is restricted to users in possession of a certificate issued by the National Authority for Medicolegal Affairs. Digital signature is used for user identification. All access is logged. Even if access right issues are sorted out, there is no guarantee that EPR systems are infallible, entry error is certainly possible. Also, deficient software reliability can fail a well-designed EPR system. ERPs that depend on certain operating systems (OS) may let down the entire system due to software bugs (Cohen, 2005).

There are several potential issues, as we shall see when we use the case study in the Dutch example (Spaink, 2005). The general public does not seem to endorse the idea of the Civil Service Number (CSN) system being implemented uniformly among all government services including healthcare, law enforcement, education, and taxation. It is reported that such a system, without accompanied EPR software support for Dutch citizens, is used primarily for promoting biometric identity cards rather than letting citizens view their EPRs. Consequently, it turns out that benefits brought to patients by EPR are not widely accepted. Although EPR is promised to bring benefits such as reduction in medical errors and costs, and less healthcare bureaucracy is involved in the national health system, its implementation may require benefits brought to users to be clearly identified.

Throughout the world, issues such as entry error and privacy issues are major factors that discourage public acceptance. Authorities need more comprehensive plans and public education well before EPR rollout. Although technologies that support EPR make it more accessible, it may still be some time away from being taken up by the general public.

6.4 Anonymous Data Collection and Processing

As we discussed in the Finland EPR system an individual takes up 100 GB of data storage for one's medical record. Comparing this to 250 GB that a typical laptop computer manufactured in 2010 has, we see that one single person's medical history entails a vast amount of information. Throughout the book we have looked at many aspects of health information, these include vital signs, images, records of each doctor visit, and medications taken; essentially everything related to health that begins from a person's birth throughout their entire life. Such information gives a lot of details about an individual.

The vast range of information types make medical records extremely useful in many areas: marketing, government planning, and pathology analysis alike. Companies utilize plentiful resources to find out the state of individuals for the purpose of segmentation marketing.

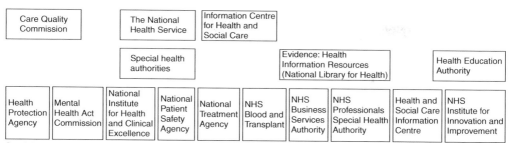

Special health authorities: Health Protection Agency; Mental Health Act Commission; National Institute for Health and Clinical Excellence; National Patient Safety Agency; National Treatment Agency; NHS Blood and Transplant; NHS Business Services Authority; NHS Professionals Special Health Authority; Health and Social Care Information Centre; NHS Institute For Innovation and Improvement

Figure 6.12   Healthcare service infrastructure

Although marketing is an important topic in promoting healthcare technology-related services, it is not within the scope of this text and readers are encouraged to read Hung (2009) for details.

We shall look behind the scene at matters directly related to national healthcare services. Serving everyone in an entire country and across all healthcare needs is not an easy task and involves very complex governing structure. For example, in the UK alone the following authorities exist, just to name a few: NHS (including separate entities for NHS Wales, Scotland, and Northern Ireland), 10 Strategic Health Authorities serving different areas, the Health Education Authority, and 10 Special Health Authorities responsible for different health-related issues. The healthcare structure is simplified in Figure 6.12. In addition to tens of authorities in England alone, there are also hundreds of trusts throughout the UK, these cover areas such as ambulance, mental health, and numerous primary care trust (PCTs). With so many entities employing over one million people throughout the healthcare system, who can access what information and how data can be shared remains a complicated situation. In case any detrimental event of information leaks out, mechanisms must be available for finding out what has happened and any employees accountable for the event can be tracked.

6.4.1   Information Sharing Between Different Authorities and Agencies

Before the IT era when paper was the main medium for circulation, information sharing between authorities and agencies took place in a very limited manner on a case by case basis. Back in the 1980s medical information sharing was not supported in real-time (Thacker, 1983). Data that flows between different entities such as clinics, hospitals, laboratories and insurance companies in the provision of healthcare services has commonly been shared at an aggregated level. Data sharing is guided and restricted by both excessive legal regulations and derisorily written guidelines. Many hurdles must be cleared in order to establish a proper process of sharing information between agencies. Most of these hurdles are not related to technical issues. Political and social hurdles are difficult but important to address. In many countries, bureaucracies as well as differences in state and federal laws may greatly hinder the prospects of establishing such a process greatly. To illustrate the establishment of this process, we look at an example with the case study in Zeng (2004).

The elevated importance of bioterrorism as a threat to public health and safety stipulates the development of disease surveillance and information sharing mechanisms for supporting real-time data analysis, and circulating information about outbreaks, both for naturally occurring and manmade viruses (Clinton, 1999). As EPRs transform from paper-based log cards to electronic database management systems, issues such as interoperability, flexibility, accessibility and scalability become increasingly important to consider. The West Nile Virus-Botulism is an example of a national infectious disease information infrastructure designed for capturing, accessing, analyzing, and visualizing disease-related information from various sources to support real-time reporting and alerting functions. The system is so vast that it contains data from human, different types of animals and insects that can potentially carry diseases, as well as botulism data. This cover involves many agencies associated with public health and safety, animal and pest control, and the National Institute of Allergy and Infectious Diseases that is responsible for botulinus intoxication in the US. Tracking of Clostridium botulinum bacteria, the bacterium that causes infantile botulism, is jointly accomplished by all participating agencies. This involves a large number of agencies because the bacteria affects people who have eaten improperly prepared raw or parboiled meats and is linked to contamination. As such, there are many possible sources. Other related information such as climate pattern and bird migration is also recorded for analyzing and tracking disease occurrence and spread. In the US, state and local regulations govern information sharing between agencies that may require prior approval from governing hierarchy of the agencies involved, prohibiting informal information sharing agreements between those agencies. These regulations may differ in terms of confidentiality requirements and duration that data can be kept for. Due to privacy issues, certain regulations may prohibit unique identification of individual persons or actual locations, making disease tracking more difficult.

6.4.2 Disease Control

Disease outbreaks can spread across states, countries, and continents. Outbreak can be *overt* and *covert* in nature, meaning it can be easily observed as a natural cause or covert as surreptitious on purpose. Overt outbreaks may be initially noticed and managed by a public health agency; whereas covert outbreaks may not initially be so familiar and hence initially managed by a public health agency then passed on to a law enforcement agency for tracking the source (Butler, 2002). Overt diseases can often be tracked initially through a geographic spread by animal or human movement. The possibility of a covert disease outbreak triggered by large-scale bioterrorism events whose occurrence may appear random, such that information sharing may involve national and international entities, the process of establishing data sharing mechanisms may be far more complicated than our above case study. Analysis of spread pattern, for both natural virus and bioterrorism, entails collection of data about the time and place of occurrence for each reported case so that a quantitative model can be constructed.

In epidemiology, disease spread usually exhibits a certain spatial pattern over time. Spatial and temporal dynamics of a virus during an epidemic is usually examined to predict the rate of virus spread. A long term study by (Viboud, 2006) has shown that virus outbreaks exhibit hierarchical spatial spread evidenced by higher pairwise synchrony between areas of higher population densities through quantifying long-range dissemination of infectious diseases. A statistical model that describes the spatial spread pattern can computed by a number of

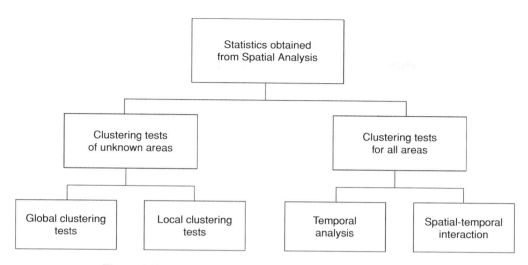

Figure 6.13   Process for infectious disease spread pattern analysis

methods, such as that proposed by Pasma (2008), the process is simplified in Figure 6.13. This chart shows the sequence of manipulating spatial analysis statistics by performing some general clustering tests based on the principle of whether the clustering location of a disease is known or not (Besag, 1991). A cluster is defined as closely grouped cases of disease with a well-defined spread pattern over space and/or time (Ping, 2006). Clustering tests of unknown locations are divided into global and local tests. The former is useful in determining any relationship between reported cases and clustering exists if cases occur in such close proximity that they spread out of control, such as the initial outbreak of A(H1N1) swine influenza in Mexico (AVMA, 2009); the latter, local clustering tests show risen rates of cases and identify affected areas as clusters by comparing rates of diseases in different clusters. Clustering tests are also performed for all areas in order to draw a clearer picture about the spread pattern. Temporal analysis helps the detection of clusters of disease over time and spatial-temporal interaction analysis uses both space and time information on cases within proximity that occurs around the same time.

Cluster analysis and detection involves detection of clustering of disease in historical disease data, focused cluster analysis, and spatial cluster detection. The main purpose of deducing a clustered spatial structure is to group what may otherwise appear to be randomized cases of diseases together to reveal a pattern with data description for visualization. Apart from spatial and temporal information that can be used to build a statistical model for predicting the pattern of disease spread, other useful information for containing and controlling disease includes demographic factors of infected persons and in cases of evaluating the severity of an epidemic medical history about pre-existing conditions and prescribed medications are also extremely useful.

Accurate prediction of disease spread as well as balancing between data capturing and respecting patient privacy are equally important in disease control. Let us take a look at one single case of swine flu that put over 300 healthy people into enforced detention for one week (Yuan, 2009), where the local authority decided to quarantine all 340 hotel guests and employees as a result of one guest having been diagnosed with the A(H1N1) virus that accumulated 10 000 cases in just over four months afterwards with a mortality rate that is statistically similar to that of most other influenza strains. While we are not in the position of debating whether the compulsory detention was justified, let us take a look at some facts gained over the five months after the incident so that we can get a good understanding about what prediction of virus spread pattern can do for us and what needs to be compromised between public health and privacy:

- Comparing the H1N1 swine flu with H5N1 avian flu in the same vicinity, H5N1 is statistically much more deadly, whereas H1N1 exhibits a faster rate of spread.
- H5N1/H1N1 recombinant virus poses a serious hazard yet there is no scientific proof of such occurrence.
- In the 1918 flu pandemic, the 'Spanish flu' that spread to virtually anywhere on earth, A(H1N1) was blamed as deadly virus that killed no less than 50 million people, mainly healthy young adults (Mitka, 2005).
- Antiviral drugs such as Tamiflu™ are adequately effective with only isolated cases of drug resistance (Community Central, 2009).
- Fairly high prevalence with more than one in 1 000 infected within four months of the first reported outbreak, carriers can travel in and out undetected by planes, ships, and trains.
- Regular seasonal flu virus, H3N2, possess greater threats than H1N1 (Higgins, 2009)
- A second strain of A(H1N1) may have mutated around the same time of the incident (Fox, 2009)

Further to various attributes of the virus itself, there are also humanity issues:

- Hundreds of tourists have their vacations withdrawn.
- Sudden loss of business opportunities without proper planning can cause a wide range of problems (Cheng, 2008), which affects both the hotel itself and guests business trips.
- People are involuntarily detained in a confined room for one week, for example, (Weaver, 2009) reported a prostitute was made to live with her customer together sharing the same room day and night, as she was not given a separate room.
- Logistics support from over 300, from basic necessities to entertainment.
- Businesses surrounding the hotels suffer as the area was cordoned off.

Was this a publicity show, was there a genuine need, or was there any political agenda? To answer this question, we look at some clues from a dozen points listed above. Digging deeper into the story, we notice that if the authority had known earlier that the virus would still spread throughout the entire world irrespective of how this case was handled, they would probably have sent the only patient into hospital while letting the remaining 300 people live as usual. In the end, this is back to the knowledge of disease spread pattern.

More knowledge about the disease would also provide authorities with necessary information for prevention and diagnosis. The swine flu virus may increase acidity of blood, as Figure 6.14 shows the blood sample of a patient with A(H1N1) magnified by 300 times. This sample is likely to suggest chronic fatigue syndrome for the patient. It may also appear similar to a mycoplasma infection, which also looks like pneumonia or SARS (He, 2003). Such symptoms may suggest urgent respiratory treatment is required.

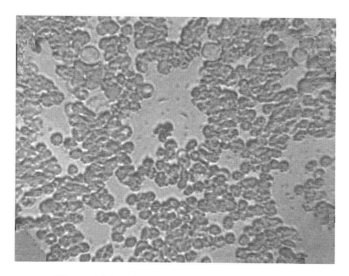

Figure 6.14   Blood sample of a swine flu patient

Now we understand the importance of finding an accurate prediction of how a virus may behave when it spreads. This is often much easier said than done. Tracking the space and time changes of the spreading process may depend mainly on anonymous information about each reported case. Gaining a more comprehensive picture of the disease may, however, involve analyzing information of individual patients, including where and when the patient has been, with whom they have has been in contact, and even the transportation mode of a journey, as (Yuan, 2009) reported even taxi drivers were caught up in the incident in the above case study. Acquisition of personal information may affect policy planning, as we shall discuss in the next sub-section.

6.4.3   Policy Planning

Authorities often collect information for planning. This involves many different agencies and departments for matters related to education, prevention, healthcare system infrastructure, and emergency services. Traditionally, statistics related to specific types of accidents have been used by authorities so that they can design education campaigns for certain groups of people who statistically have higher risks, for example, statistics about alcohol-related traffic accidents have been analyzed to find information about times and locations of likely occurrence and campaigns are set out to target those at higher risk of drink driving. This seemingly simple course of combating drink driving involves many entities. In the UK, the Department of Transport organizes campaigns and promotional materials for accident prevention, conducts surveys of attitudes to road safety and provides related information; local drug and alcohol departments may also assist with rolling out campaigns. Of course, the police are out to catch the drink drivers. The Royal Society for the Prevention of Accidents (RoSPA) takes care of driving safety and training; and the Campaign Against Drinking & Driving (CADD) is established as a charity to support crash victims and their families. Each of these entities has its own specific functions and they collect data from different ways for different purposes. They all share one common responsibility: keeping data with strict confidence and ensuring that data will not be lost or stolen, regardless of how the data is treated.

Statistics are gathered and some shared by various entities. Obviously, data collected cannot be sent 'as is' because it may contain sensitive information. For example, a result collected from a breath test is associated with a specific driver, whose detailed information including address, driving licence number and vehicle registration details are all gathered. Information shared for statistical analysis may include time and place of the test, age group of the driver, and the breath test result. Nothing that can uniquely identify the driver being tested should be transferred from one entity to another without the driver's consent.

Health statistics for policy planning sometimes involves agencies of a certain region such as a county or a province, or at a national level that handles statistics from all areas within the country where each local government may have its own agencies with different functional duties for planning. At the same time, they also interact with national agencies in a broader context. For example, each state has its own agency with different functions in the US. In our case study, we look at the structure of two adjacent states in the western US, California and Nevada. California has its Center for Health Statistics (CHS), whose primary function is to manage the collection and distribution of health related statistical data. It consists of several offices and sections as shown in Figure 6.15. These offices are each responsible for a number of functions such as maintenance of a system for registration for all births, deaths, and marriages; including issuance of appropriate certificates, vital records also include over one million miscellaneous incidents each year. Health information gathered is used for research on

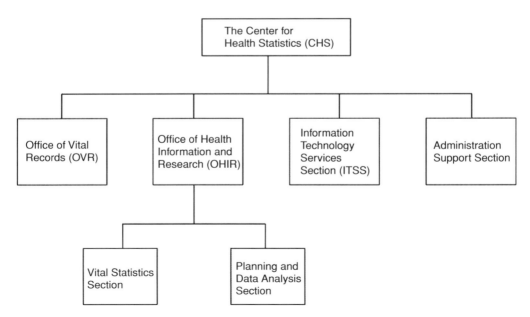

Figure 6.15   Healthcare service organizational structure

general health status of the state's residents. California's CHS is also responsible for disease control, local health services, and regulation of the public drinking water systems where the vast geographical coverage of the state denotes the need for different field operations branches for the northern and southern parts of the state, in conjunction with the Technical Operations Sections, Monitoring and Evaluation Unit, and the Infrastructure Financing & Infrastructure Funding Administration Sections. Within the California state, health data including information about individual residents are also used together with agencies such as the Department of Health Care Services and the California Health and Human Services Agency.

In the state of Nevada, health information is administered by the Bureau of Health Statistics, Planning & Emergency Response (HSPER), under Nevada's Department of Health and Human Services. It too is responsible for birth, death, and marriage registration, vital and health statistics analysis, and public health monitoring. Unlike California where the Office of Statewide Health Planning and Development, under the California Health and Human Services Agency, is responsible for the health planning; health planning in Nevada is also handled by the HSPER. Public health planning is extremely important in ensuring optimal utilization of available resources and residents are well cared for.

Health planning, irrespective of organizational structure, involves the collection, storage, processing, validation, analysis, and distribution of health and related data. Such data allows authorities to plan for future healthcare services such as prediction of future needs and preventive education. For example, using information about population growth and statistics on usage of various units of surrounding hospitals would enable planning for a new hospital with an appropriate size of each unit in it.

Many countries conduct a census for prediction and planning for a wide variety of needs. These are usually carried out by a government agency dedicated to collecting and analyzing data once every few years (United Nations recommendation is once every ten years). The process of acquiring information about the population dates back to the eleventh century when the 'Domesday Book' of the year 1086 appeared in England with information about individual households. Census is now conducted by the Office for National Statistics (ONS) in the UK for planning, including healthcare policies. In the US, this is handled by the Census Bureau whose functions are similar to UK's ONS.

Another piece of important information that Census gives is aging population. This growing problem in most developed countries means demand for healthcare will rise over the next two to three decades. A snapshot of the UK's National Statistics Online in Figure 6.16 shows a *population pyramid*, it conveys some basic information that is helpful for long-term healthcare policy planning.

1. A clear spike around the age of 60 results from the post-war baby boom that tells a large group of people are around the retirement age, over the next decade these people will very likely need more healthcare assistance.
2. There was another baby boom around the mid-1960s, now around the early to mid-40s. We can deduce that an influx of retirees will utilize the healthcare system in about 20 years' time.
3. It also gives information about migration, a net increase in migration (i.e. the number of immigrants settling down exceeds that of the number of people emigrating from the UK to

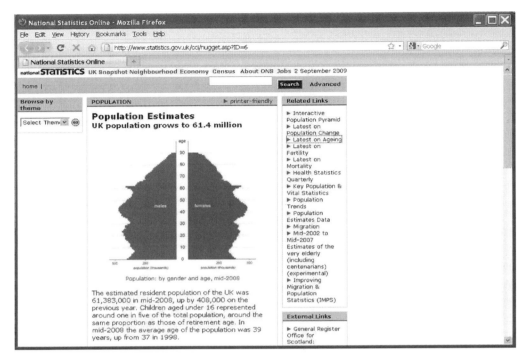

Figure 6.16   Screen shot of a population pyramid

settle permanently overseas). Immigrants may settle in certain areas within the countries. Therefore, an increase in demand may suddenly affect individual hospitals.

4. This pyramid shape reflects a similarity in number between both genders among the UK population. The balance shifts towards the female side from around the age of 70, which is in line with the fact that life expectancy of women is higher than that of men. Elderly women and men may have different chronic diseases that require different treatment. Such trends can be used to estimate the needs.

5. By comparing population pyramids of different years, an upward trend confirms population aging is becoming a greater problem. Such study allows authorities ample time to develop policies in anticipation to the growth in healthcare demand.

It is worth noting that the process of analyzing population statistics is a very lengthy course that may take more than one year. The information in Figure 6.16 pulled out from National Statistics Online was released at the end of August 2009, which shows data for mid-2008. The lack of timely information and the complexity involved in statistical analysis together explain why best effort healthcare policies adopted throughout the world may not provide the best solution to the nation's needs. No matter whether the data collected is important or not, the single most important point to observe is privacy issues when collecting data.

6.5   Biometric Security and Identification

Biometric refers to some form of physical characteristics measurement of the human body. Such a unique feature that belongs to an individual can be used for security and identification purposes. A person's voice has been used for recognition for decades, not very conclusive, although speech recognition and filtering algorithms have improved notably over recent years. Although biometric security is used in many areas outside the medical domain, it is a topic that warrants noteworthy discussions owing to its popularity in different areas of healthcare and telemedicine related applications.

French anthropologist Alphonse Bertillon (Ferembach, 1989) was perhaps the first person who formally documented biometric identification. His pioneering work created *anthropometry*, a systematic biometric measurement for unique personal identification. Anthropometry is accomplished by some kind of measurements of certain body parts. The original anthropometrical system of identification consisted of three parts:

1. Physical Measurement:
   Precision measurement under some stipulated conditions, measurement includes characteristic dimensions of parts of a human body; for example, size of the ear.
2. Morphological Description:
   The shape and contour of the body which entails a characteristic description of mental and moral attributes, something that relates to the movement of a certain part such as limbs.
3. Peculiar Marks Description
   Observation of any signs anywhere on the body that may be left from an accident, disease, or disfigurement. These include scars, moles, and tattoos.

Such a complicated process proposed by Bertillon may even result in two sets of different results being obtained from one individual when the process is repeated. To simplify Bertillon's identification process, it was later said that the ear was sufficiently unique such that he quoted 'It is, in fact, almost impossible to meet with two ears which are identical in all their parts.' (McClaughry, 1896). It described a series of four visual features of an ear as follows:

1. 'Helix': three portions of the border of the ear, and the extent of openness
2. 'Contour': degree of adherence to the cheeks and the ear lobe size
3. 'Profile': inclination from horizontal and the amount of reversion in front of the antitragus
4. 'Fold': dimension and pattern of anthelix (anteroinferior to the helix)

The details documented were so explicit that McClaughry (1896) gave a 15-page description of every single feature of the ear, including the earlobe, the tragus, the antitragus, the concha, and the superior fold. Such in-depth description of the ear, although has never been known to be applied to personal identification, did, however, serve as an important milestone for systematic identification by bodily measurements.

One obvious application of biometric identification is restricting access to different parts of a hospital; where staff members can enter an area by being uniquely identified without the risk of lending out one's identity card. Cases of lost or stolen cards have been heard from time to time. Using identification with biometric authentication would certainly ease such a problem. Apart from behavioural uniqueness such as signature and voice of an individual person, there are certain physiological attributes of a human body that are unique in nature with practically zero chance of finding two persons, even twins, who possess identical properties. These include finger and palm prints, iris patterns, facial patterns in terms of optical and thermal outlines, and deoxyribonucleic acid (DNA). Whereas DNA is not commonly used because current technology still requires some form of tissue to be analyzed.

6.5.1   Fingerprint

Fingerprint identification is perhaps the most well-known method of biometrically identifying a person. A fingerprint impression consists of patterns that exhibit physical differences between ridges and valleys on the finger's surface (Lee, 2001); where ridges and valleys refer to the upper and lower segments of the skin, respectively. Where each ridge ends, it forms a *minutia point*, where the size and shape can also differ. Also, a *ridge bifurcation* refers to where a ridge splits into two as if branching out. The positions of these unique features as shown in Figure 6.17 are used to identify a person. Ridges and valleys are show in black and white, respectively. There are five fundamental fingerprint patterns, namely: whorl, arch, tented arch, left loop and right loop. Generally, loops cover about 2/3 of the fingerprint and whorls cover about a quarter of the finger. The remaining 10% is covered by arches. In a loop, one or more of the ridges enters on either side of the impression. A loop consists of a *core* and a *delta*, these are circular and triangular patterns that when grouped together form a loop. In a whorl, some of the ridges turn several times. A portion with at least two deltas is considered to exhibit a whorl pattern. Classification can generally be accomplished by counting the number of deltas.

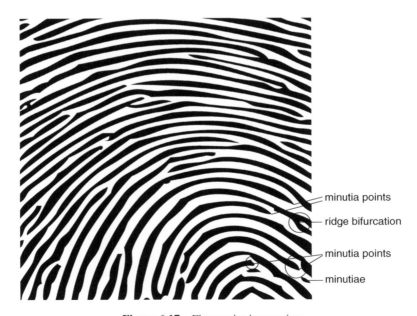

Figure 6.17   Fingerprint impression

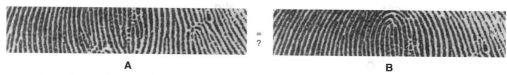

A	B
Impression stored on record for authenication	Scanned image of the same finger but of different portion

Figure 6.18   Scanned image of a portion of a finger under different alignment

Lack of delta is an arch, one delta is a loop, whereas at least two deltas make a whorl. In an arch, the ridges run from one side to the other across the pattern.

For more than a century, fingerprints have been widely accepted as an infallible means of unique identification. Fingerprint analysis has put countless criminals behind bars everywhere throughout the world, this is a proven technology based on the comprehension that no two fingerprints have ever been found identical among billions of people who have ever existed. The study of fingerprint impression was first formally studied by Czech physiologist Jan Evangelista Purkyně (Purkyně, 1823), after earlier work that studied skin's ridges in Grew (1686), where he illustrated nine fingerprint patterns that were later used for identification purpose.

Powerful image processing algorithms and lower cost of high resolution scanners make such applications so popular that fingerprint identification has been seen in many consumer electronic devices, many medium to high end laptop computers now include a narrow strip of optical scanner that scans a portion of the user's fingerprint. When used in this kind of consumer electronics devices, it relies on comparing the portion of scanned fingerprint with the impression that has previously been stored. Since repetitive scanning of the exact same portion of the finger during different occasions is practically impossible, authentication is only done by using a small portion of the finger, as illustrated in Figure 6.18. Compromise is made between the physical size of the optical scanner (where the user puts only a portion of the finger on) due to space saving and manufacturing cost reduction, verses the ability to uniquely identify an individual person. When we look at Figure 6.18, the impression in (a) was originally stored as reference. When the same finger is scanned again at a later time in (b); with reference to the stored impression, the finger was placed a little further to the right, and somewhat below. So, only a fraction of the image in (b) is identical to that of (a). As the stored reference does not contain the entire finger's impression, the authentication algorithm needs to extract a certain portion of the scanned fingerprint image in order to make a comparison to the reference. In this particular example, there is a sufficiently large portion that 'overlaps', so that authentication is successfully performed when comparing the two images. There may also be circumstances where alignment is so far off that the scanned portion would appear too different, such as that scanned in Figure 6.19, which shows a lower portion of the same finger, almost the entire scanned image is out of range relative to the stored reference.

6.5.2   *Palmprint*

In addition to palmprint patterns on the skin that exhibit certain visually observable similarities with fingerprints, veins inside the palm also have unique patterns that can distinguish an individual person. Radiation of near-infrared rays can construct an image with the differing

Figure 6.19   Another scanned portion of a finger

absorption rate by deoxidized haemoglobin in the palm vein. So, the veins will appear as black lines due to less reflection of the rays. The main advantage of scanning vein pattern is that the user does not have to touch the scanner during the image capturing process so that the scanning process can be faster.

The earliest recorded case of printing human hands and feet impressions was during the pyramid construction era in Egypt some 4 000 years ago. Long before this, a small portion of palmprint has been reportedly found in Egypt that dates back to 10 000 years with an impression on hardened mud. In the modern world, there are situations where palmprint scan is more convenient than using fingerprints, particularly in telemedicine applications where the entire hand is involved in operating a system as intervention by the user is minimal. Palmprint is more appropriate in situations such as robotic surgery involving haptic sensing, where the surgeon can be identified while performing an operation. Also, in telecare systems where elderly patients move around, logging of exit and entry by placing the entire hand on the scanner would be faster and easily than to align one finger on a fingerprint scanner.

Identification of a palmprint is usually accomplished by local feature extraction through a voting scheme that combines a set of fuzzy k-nearest-neighbour (k-NN) classifiers (Hennings-Yeomans, 2007). Image preprocessing is performed with global histogram equalization for the scanned image of size $M \times N$ with G grey levels and cumulative histogram $H(g)$, whose transfer function is:

$$T(g) = \frac{(G-1) \times H(g)}{M \times N} \tag{6.1}$$

The local histogram equalization is then applied by cropping the image starting in the upper-left corner having a pre-determined window size, followed by applying the histogram equalization function to the cropped image. The same process is then repeated by moving the crop all over the image and for each one applying the equalization. This mathematical description of the scanning process may sound fairly complicated, the process itself is actually quite simple: first, the palm impression is scanned to generate a monochrome image of the palm and certain features throughout the palm are identified and extracted, a certain portion of size $M \times N$ number of pixels is extracted and represented by varying levels of grey shades. A repetitive equalization process is performed progressively throughout the image, starting from the top left corner of the impression.

Similar to fingerprint authentication, the user to be authenticated has a digital image of the palm captured from a scanner stored as a reference. The scanner resolution must be high enough so that palm lines are detectable in subsequent processing stages and image analysis. Also, the background of the palm image should be as plain as possible without any pattern that may be misread as palm lines for the convenience of edge detection. Prior to the authentication step, the background information should be removed by the image processing algorithm.

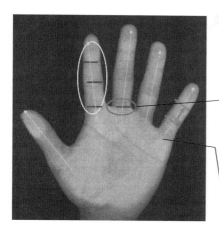

1: Find all line segments separating all fingers.
2: Group all parallel line segments. Five groups are expected as a result. Assume that a normal palm is examined, thus each group contains 3 line segments* except one group for thumb that contains only 2 lines.
*It is possible that for some people, there might be 3 separating palm lines for thumb and 2 separating palm lines for the rest of the fingers.
• Note: Also notice that more than 1 separating line segments on a finger will be detected. A threshold value of distance will be determined to group all line segments near to each other, and an <u>average line segment</u> will be calculated.
3: After finding all separating line segments of all fingers, a line normal to one group and passing the mid point of all line segments will be determined. The one which is normal to one group containing 2 line segments only is the one on the thumb. All other normal lines of various fingers can be determined according to the angle between them and the normal line of thumb. After we have the 5 normal lines and all separating line segments, the <u>length of each section of a finger</u> can be easily calculated.

Details on finding all line segments:
1. Segments must be of a certain length in order to be considered. Segments that are too short will not be considered while doing averaging to avoid inaccuracy.
2. Segments must also not be too long and must have some parallel mates, to avoid detecting line segments at the center of palm.
*Threshold values will guide through the process.

Figure 6.20 Palmprint scanning

The process first involves palm alignment. To the human eye identifying where the palm is may sound easy. To machinery where optical sensors attempt to identify the palm, it must first identify where the palm is placed. This involves removal of any background information that does not represent any part of the palm. Having removed the background, the position of the palm will be identified. Once the position of the palm is identified, sub images of palm centre and fingers can be transformed from the original palm image using co-ordinate transformation techniques in the next module: palm partition. Since finger length and lengths of parts of a finger are parameters of great importance in finger analysis, a method of finding line segments is described in Figure 6.20.

When data is sent across a telemedicine network, there will be a trade off between power usage versus palmprint recognition which is compared to evaluate the percentage of power necessary to achieve an acceptable rate of scan so that the scanner's power consumption can be minimized while maintaining adequate performance.

6.5.3 Iris and Retina

Eye scan depends on the unique pattern of blood vessels, retina scanning is accomplished by shinning a low intensity light beam into the eye and the reflected light generates a pattern of the *capillary* (the minute connections between arteries and veins) in the retina. The blood vessels on the retina absorb more light than the surrounding tissues therefore a monochrome image of different grey shades representing the darker blood vessels and the brighter background can be formed. A scanner is normally placed around one to two centimeters (or 0.5 in.) away from the eye, light that reflects back from the retina will form an image revealing what looks like the illustration in Figure 6.21. The image, captured by the retina scanner, exhibits a unique pattern

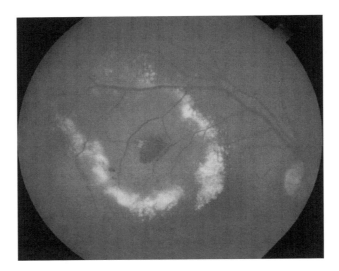

Figure 6.21   Image of the retina

of lines. Figure 6.22 shows the retina scanner used in acquiring the image in Figure 6.21, it may look fairly similar to a webcam but its internal architecture and functions are very different.

As blood vessels run within an eye in a 3-D space, a scanned image that effectively produces a 2-D representation may lose certain details that compose the overall impression of the eye. Such loss in image quality is unlikely to be as serious as any changes due to diseases

Figure 6.22   Retina scanner

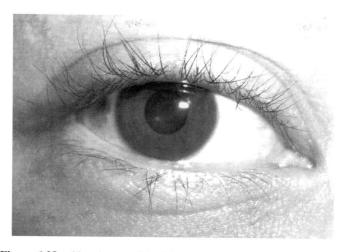

Figure 6.23   Clear image of the iris under the influence of contact lens

such as cataracts, diabetes and glaucoma. Owing to the high resolution details necessary for identification, retina scanning only became popular over the past couple of decades when affordable 3-D scanning became possible despite its scientific concept being founded as early as the 1930s (Simon, 1935). Although retina scanning is traditionally used by government agencies for authentication and more recently expanded to civilian use, one of its important medical uses is early disease detection as potentially fatal illnesses such as chicken pox, malaria, and sickle cell anemia; as these conditions affect the eye even during very early stages of prevalence. Similarly, chronic diseases like atherosclerosis also affect the eye at an early stage.

Another ocular scanning technique is iris scanning. Instead of using veins, iris scanning uses its intricate structures, which is widely accepted as a unique feature of the eye. One major convenience is that its effectiveness is not affected by glasses or contact lenses. The iris features in Figure 6.23 can still be clearly seen despite the subject wearing a contact lens. With reference of a scanned photographic image of the subject's eye, it operates by first extracting the boundaries of the iris and the pupil. This identifies the portion of the image that forms part of the iris. Image recognition is performed with part of an eye because the iris is normally partially covered by eyelids therefore it would be impractical to make use of an image of the entire eye. Less restrictive than retinal scan where the subject needs to align an eye very close to the scanner, iris scanning can normally be performed around one metre (or 3 ft) away, so the subject does not have to sit right in front of the scanner.

Iris scanning is becoming increasingly popular as the related optical technologies become more mature. The UK Border Agency runs the Iris recognition immigration system (IRIS) at major international airports in England, so that registered persons can enter the UK by looking at installed cameras while walking through the immigration arrival hall. At the time of publication, information about registration and technical details are available from the UK Home Office website at: http://www.ukba.homeoffice.gov.uk/managingborders/technology/iris/

Eye scanning, although widely considered as a more accurate method than using fingerprints, does have a major problem as it is considered as unsuitable for frequent scanning. Prolonged exposure to light emitted by the scanner can be harmful to the eye hence frequent use is not recommended. Although both techniques entail scanning of typically around 10 - 20 seconds, iris scanning would be less harmful due to the distance separation between the eye and the scanner.

6.5.4 Face

Not to be confused with the face detection function on many modern digital cameras, where algorithms are designed to recognize certain area within a scene as a human face. Face recognition uses certain features of an individual person for unique identification using techniques in computer vision and image processing. Unlike the three methods described above that principally involve comparison with a reference still image stored on record, face recognition also works with video (collection of constantly changing images at a certain frame rate per second). Recent development in facial recognition algorithms allow 3-D representation of the shape of a face, such that contours of eyes, nose, lips, and the chin can be expressed in digital format. While most of the methods described above use some kind of optical sensors that require the subject to remain stationary for a short period of time, with 3-D expression recognition can be accomplished from different angles (Bonsor, 2008). In addition to face shape, skin texture with details such as patterns and unique visual features can also be used together for a more relevant way to describe a subject's face.

This technique is already in use by the US Department of State where photographs for entry visa application are store for facial recognition of each applicant. In year 2000, London Borough of Newham in the UK also implemented its *FaceIt* surveillance system in an attempt to track down criminals who are caught in CCTVs (Closed Circuit Televisions). By 2006, police in England and Wales have expanded the use of face database and set up the *FIND* (Facial Images National Database) for identifying criminals. All currently deployed systems not only face technical challenges, privacy is also a major issue especially when CCTVs are deployed.

Following the first textbook exclusively on face recognition by (Hallinan, 1999), several volumes on face recognition have been published in the early 2000s, covering a wide range of topics related to recognition from facial image and the effects of demographic factors, there are also suggestions that the subject's gender may also contribute to the effectiveness of face recognition. For further reference on face recognition and technical details, readers are encouraged to refer to Delac (2008); a free electronic copy in a zip file of size 16.7 MB is available for download from this link as of March 2010: http://intechweb.org/book.php?id=101

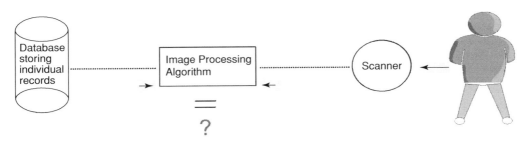

Figure 6.24 Framework for user identification over a telemedicine network

This chapter has discussed a number of areas in information technology where patient information can be safeguarded. We conclude by taking a look at Figure 6.24 where a generalized block diagram shows how various biometric authentication techniques can be linked up to a telemedicine network for identification for user access as well as telecare applications. For all the techniques described in section 6.5, we have a database that stores information about each individual person, comparison from an image obtained from a scanner with reference to each record will determine a subject's identify.

References

AVMA (2009), American Veterinary Medical Association (AVMA): 2009 H1N1 Flu Virus Outbreak http://www. avma.org/public_health/influenza/new_virus/default.asp

Barron, H. K., et al. (2007), The Electronic Patient Record, House of Commons Health Committee- Sixth Report of Session 2006-07 Vol.1, published by authority of the House of Commons, London. The Stationery Office Ltd., 13 September, 2007.

Besag, J. and Newell, J. (1991), The detection of clusters in rare diseases, *Journal of the Royal Statistical Society: Series A*, 154:143–155.

Bonsor, K. and Johnson, R., How Facial Recognition Systems Work, *"Online article"*: http://electronics. howstuffworks.com/gadgets/high-tech-gadgets/facial-recognition.htm, October 30, 2008.

Butler, J. L., Mitchell, C., Friedman, C. R., Scripp, R. M. and Watz, C. G. (2002), Collaboration between public health and law enforcement: new paradigms and partnerships for bioterrorism planning and response. *Emerging Infectious Diseases*, 8(100):1152–1156.

Cheng, L., Leung, T. Y. and Wong, Y. H. (2008), *Financial Planning & Wealth Management: An International Perspective*, McGraw Hill.

Clinton, W. J. (1999), Remarks by the President on keeping America secure for the 21st Century. National Academy of Sciences, Washington, D.C., January 22, 1999.

Cohen, A. (2005), Software Is Too Buggy and Unreliable, *PC Magazine*, 3 August, 2005.

Community Central (2009): H1N1 Tamiflu resistance concerns examined http://www.continuitycentral.com/news04661.html

ComputerWorld Hong Kong Staff (anonymous) (2009), United Christian Hospital loses another USB drive, *ComputerWorld April 15, 2009*.

Delac, K., Grgic, M., and Bartlett, M. S. (2008), *Recent Advances in Face Recognition*, In-Tech, ISBN: 978-953-7619-34-3.

Ferembach, D. (1989), Alphonse Bertillon (1853–1914), *International Journal of Anthropology*: 4(4):309–311.

Fox, M. (2009), Second strain of flu may complicate picture, *Reuters* report, May 6, 2009.

Greenhalgh, T., Wood, G. W., Bratan, T., Stramer, K., and Hinder, S. (2008), Patients' attitudes to the summary care record and HealthSpace: qualitative study, *British Medical Journal*, 2008,336:1290–1295 (7 June).

Hallinan, P. W., Gordon, G., Yuille, A. L., Giblin, P., and Mumford, D. (1999), *Two- and Three-Dimensional Patterns of the Face*, A K Peters Ltd., ISBN: 978-1-56881-087-4.

He, M. L., et al. (2003), Inhibition of SARS-Associated Coronavirus Infection and Replication by RNA Interference, *Journal of the American Medical Association* 290:2665–2666.

Hennings-Yeomans, P. H., Kumar, B. and Savvides, M. (2007), Palmprint Classification Using Multiple Advanced Correlation Filters and Palm-Specific Segmentation, *IEEE Trans. Information Forensics and Security*, 2(3 II):613–622.

Higgins, R., Eshaghi, A., Burton, L., Mazzulli, T., and Drews, S. (2009), Differential patterns of amantadine-resistance in influenza A (H3N2) and (H1N1) isolates in Toronto, Canada, *Journal of Clinical Virology*, 44(1):91–93.

HIPAA: http://www.hhs.gov/ocr/privacy/hipaa/understanding/index.html

Hung, H., Wong, Y. H. and Cho, V. ed., (2009), *Handbook of Research on Ubiquitous Commerce for Creating the Personalized Marketplace: Concept for Next Generation Adoption*, IGI Global, USA.

Lee, H. C. and Gaensslen, R. E. (2001), *Advances in Fingerprint Technology*, 2/e, CRC Press USA, ISBN: 9780849309236.

McClaughry, R. W. (1896), *Signaletic Instructions: Including the Theory and Practice of Anthropometrical Identification*, Kessinger Publishing, LLC (ed. October 27, 2008) ISBN: 1437270883.

Malpighi M. (1685). *De Externo Tactus Organo Anatomica Observatio* (Anatomical Observations of the External Organs of Touch). Aegidius Longus. Naples, Italy.

Mitka, M. (2005), 1918 killer flu virus reconstructed, may help prevent future outbreaks, *Journal of the American Medical Association*, 294:2416–2419.

Pasma, T. (2006), Spatial epidemiology of an H3N2 swine influenza outbreak, *Canadian Veterinary Journal*, 49(2):167–176.

Ping, Y. and Clayton, M. K. (2006), A cluster model for space-time disease counts, *Statistics in Medicine*, Wiley, Chichester, 25(5):867–881.

Purkyně, J. E. (1823), *Commentatio de examine physiologico organi visus et systematis cutanei*, Breslau, Prussia: University of Breslau.

Rindfleisch, T. C. (1997), Privacy, information technology, and health care, *Communications of the ACM*, 40(8):92–100.

Rivest, R., Shamir, A. and Adleman, L. (1978), A Method for Obtaining Digital Signatures and Public-Key Cryptosystems. *Communications of the ACM*, 21(2):120–126.

Schneier, B. (2007), *Cryptography Classics Library: Applied Cryptography, Secrets and Lies, and Practical Cryptography*, John Wiley & Sons, ISBN: 9780470226261.

Sileo, J. D. (2005), *Stolen Lives: Identity Theft Prevention Made Simple*, Da Vinci Publications, ISBN: 0977059774.

Simon, C. and Goldstein, I. (1935), A new scientific method of identification, *New York State Journal of Medicine*, 85(7):342–343.

Spaink, K. (2005), Hacking health: electronic patient records in The Netherlands, *22nd Chaos Communication Congress*.

Tange, H. (2008), Electronic patient records in the Netherlands, *Health Policy Monitor*, Bertelsmann Stiftung, Oct. 2008.

Thacker, S. B., Choi, K. and Brachman, P. S. (1983), The surveillance of infectious diseases, *Journal of the American Medical Association* 249(9):81–85.

Thomton, P. (2009), Patient Privacy; UK Law to European Standards, *A report to the National Information Governance Board for Health and Social Care*, http://www.e-health-insider.com/img/document_library0282/PT_NIGBreport.pdf

Viboud, C., Bjernstad, O. N., Smith, D. L., Simonsen, L., Miller, M. A. and Grenfell, B. T. (2006), Synchrony, waves, and spatial hierarchies in the spread of influenza, *Science* 21 April 2006, 312(5772):447–451.

Weaver, M. (2009), Quarantine brings love in the time of swine flu at Hong Kong hotel, *The Guardian* 8 May, 2009.

Weitzman, E., Kaci, L. and Mandl, K. (2009), Acceptability of a personally controlled health record in a community-based setting: implications for policy and design, *Journal of Medical Internet Research*, 11(2).

Wong, W. H. (2005), Timing Attacks on RSA: Revealing Your Secrets through the Fourth Dimension, *ACM Crossroads* 11(3):5.

Yuan, E. (2009), 1 swine flu case leads to 340 quarantines in Hong Kong, CNN news report, May 4, 2009.

Zeng, D., Chen, H., Tseng, C., Larson, C., Eidson, M., Gotham, I., Lynch, C., and Ascher, M. (2004), West Nile virus and botulism portal: a case study in infectious disease informatics, *Lecture Notes in Computer Science*, Springer Berlin/Heidelberg, ISBN: 978-3-540-22125-8. http://ai.arizona.edu/go/intranet/papers/ISI04-WNV-BOT-final%20(1).pdf

7

Information Technology in Alternative Medicine

In the US alone, the market for consumer wireless healthcare products and services are expected to soar from US\$ 300 million to 4.4 billion over the four years from 2009 to 2013 (Versel, 2009). Such growth in demand is fueled by the combined effect of telemedicine technology advances and increase in general health awareness. Consumer healthcare technology and alternative medicine will certainly become increasingly important in this regard. Alternative medicine is defined by (Bratman, 1997) as healing practice 'that does not fall within the realm of conventional medicine'. Examples such as acupuncture, biofeedback, herbal medicine, hypnosis, and yoga are collectively referred to as Complementary and Alternative Medicine (CAM). It is not difficult to realize that there is one thing in common: they do not entail drug prescription. Most of what we currently use for wireless home healthcare and maintaining fitness match the description of alternative medicine. Many of these provide us with information about our health state and possibly with suggestions of how we can improve our own health. However, almost all of these consumer health monitoring devices provide us with a direct link to medication. In the consumer electronics market, there are many healthcare related products, literately covering the body from head to toe. Some are said to improve a user's health and metabolism while others claim to keep a user in optimal shape.

On the official website of the US National Center for Complementary and Alternative Medicine, it classifies Traditional Chinese Medicine (TCM) as part of CAM. This leads to the concept of blending TCM practices such as acupressure and herbal medicine with consumer healthcare technology. With over 5 000 years of history, regulation of various aspects of a human body can help with both prevention and treatment (Yuan, 2000). Given the diverse range of benefits TCM provides, technology that complements TCM practices would certainly be of significant value to the general public. Indeed, its popularity has increased quite substantially over the past decade in the US, as the 2007 National Health Interview Survey reported a threefold increase in a decade to over 3 million Americans who utilize TCM. Obviously, TCM is only a subset of what CAM has to offer. The enormous business opportunities related to CAM warrant thorough study into how various aspects of information technology and telemedicine can make CAM more economical and viable. This chapter aims at exploring how

Telemedicine Technologies: Information Technologies in Medicine and Telehealth   Bernard Fong, A.C.M. Fong, and C.K. Li
© 2011 John Wiley & Sons, Ltd

some popular CAM remedies can be provided by consumer healthcare products and where technology comes into improving practices that have been around for millennia. Due to its vast coverage, our intention is not to go into the details of CAM. Instead, we seek to give a brief introduction to several mainstream topics inside the vast CAM market. We shall look at what technology is available for accelerating the US$ 40 billion CAM market (*Nutrition Business Journal*, 2009).

7.1 Technology for Natural Healing and Preventive Care

In unrecorded prehistoric medicine, it is generally believed that plants have long been used as healing agents on a trial and error basis. The translation of the famous *Herodotus* (Rawlinson, 1956) describes a public health system that was supported by the practice of medicine. In addition to shamanism, ancient Egyptian medicine also made use of clinical diagnostics and anatomy (Nunn, 2002). Babylonian also introduced diagnosis, prognosis, physical examination, and medical prescriptions (Horstmanshoff, 2004) leading to the evolution of modern medicine. While these focus primarily on targeting specific symptoms, ancient Chinese paid more attention to the general health state and well-being of the human body with empirical observations of behaviour forming the basis of TCM (Veith, 1972). As an alternative medicine covering regulation of the entire body from head to toe, TCM practices mainly rely on herbal medicine, acupuncture, and dietary treatment. TCM is said to rely on thorough observation of both the human body and nature (Unschuld, 2003).

Biofeedback refers to a collection of methods that relieves stress-related symptoms, and phobias. Electronic monitors are used to assist a patient gauge and response by altering the output signals. By increasing the patient's awareness of physiological activity in their muscles, one can be trained to control what are otherwise natural physical responses to tension and stress, such as heartbeat, blood pressure and breathing. The use of biofeedback intervention for treating high blood pressure has been established in clinical trials for over a decade (Nakao, 1997).

7.1.1 Acupuncture and Acupressure

Acupuncture is perhaps the most popular TCM practice widely accepted in the Western world. It relies on small areas across the anatomy associated with a specific organ or part of the body, known as *acupoints*, or *tsubo* in Japanese; there are hundreds of these acupoints scattered across the body with varying healing properties and effectiveness (Lu, 1980). Acupoints distributed throughout the body, as shown in the chart of Figure 7.1, are 'linked' to different organs or parts of the body. It is not necessarily true that an acupoint is located close to the respective organ. For example, application of a needle to an acupoint on the foot is said to be effective for relieving digestion problems. Given that so many acupoints are interconnected, its complexity makes a 2-D chart such as Figure 7.1 extremely confusing and difficult to read. It is included for the sole purpose of illustrating the chart's sophistication rather than giving any useful information about individual acupoints. Both acupressure and acupuncture rely on the flow of energy across the body through *meridians* (Maa, 2003), each meridian is a circuit that links some point on the external body to an internal organ with some kind of related physiological functions.

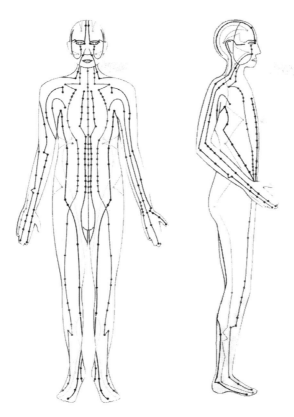

Figure 7.1   2-D acupoint chart

The aim of this book is not to discuss acupuncture/acupressure practice or properties of individual acupoints but to explore how telemedicine and related technologies can be applied to assist such practices. For this purpose, let us briefly look at some general relationships between acupoints and the human body before going into technology applications. Although we do not intend discussing any details on these practices, it is worth noting that both acupuncture and acupressure use the same points for healing; the former relies on insertion of fine needles while the latter is a non-invasive method that is stimulated with pressure exerted by a finger. In the context of our discussion on technologies that support such practices, we make no distinction between the two.

Any efforts that maintain a person's health should commence with the immune system. The concept of linking lifestyle and living longer healthily results in attention to self-care (Barrett, 1993). Its importance is gaining more attention over recent years due to the combining effect of an aging population, change in lifestyle, and stress induced by irregular work schedule among many people (Marshall, 2005). Much of CAM emphasizes balance, as in optimally balancing between various attributes for making the most of one's metabolism. The process of metabolism determines the rate at which food is digested and hence calories burnt. The basic idea is to maximize the 'efficiency' of the body by strengthening the immune system.

Stress from various daily activities causes shoulder and neck tensions as well as anxieties that affect digestion. Adding to the overall picture of 'unhealthy lifestyle' are insufficient rest and exercise that tend to make a person feel tired after a day's work. Many appliances are commercially available to provide some form of relief that partially addresses the problems caused. We shall discuss the details in section 7.4. To understand how technologies help, first we explore the effects on the immune system due to excessive overstressing of certain acupressure meridian pathways that result from prolonged or repetitive activities, commonly encountered occupational hazards such as:

- Continuous computer monitor usage: results in emotional imbalances that affect the small intestine; relieved with an acupoint located at the centre of the breastbone.
- Prolonged sitting on a chair: results in anaemia, digestive and stomach problems, relieved with an acupoint on the leg slightly below the knee.
- Excessive standing: backaches and fatigue, also causes bladder and kidney problems; various acupoints on the upper chest below the collarbone, along both sides of the spine at the lower back, and as far down as the insides of the ankles, can relief the tension caused.
- Physical exertion: causes cramps and spasms that can eventually lead to liver failure, an acupoint on top of the feet will ease the problems caused.

The above are four among countless examples where acupressure can improve a person's well-being. This is accomplished by applying firm and steady pressure onto the appropriate acupoints and can be done on a daily basis just like a regular workout (Abreo, 2009).

7.1.2   Body Contour and Acupoints

To facilitate the application of acupressure, any automated system needs to identify each apposite acupoint for a specific purpose. A reference chart only provides some indication on roughly where an acupoint is, physically locating it is far more difficult for an inexperienced person, and even more difficult for machines. Individual body size and shape can be very different, for example, the location of a specific point on a 5' tall thin person can be very different to that on a 6' tall fat person. What appears to a human eye can be perceived very differently by a machine. Further, the body contour can vary significantly from person to person. Some appliances, such as a massage chair, automatically searches for the approximate location of acupoints (pinpointing would be virtually impossible due to the precision involved) by first scanning across the user's back to obtain the related positions of the neck and spine. Several reference points can be established by visually aligning the anterior superior iliac spines (ASIS) and posterior superior iliac spines (PSIS) to vertical from the side as shown in Figure 7.2, similar to the technique used in (Fong, 2010) and can generate a graphical representation of the user's body profile such as that shown in Figure 7.3. This sample body profile shows the sitting position of a user. Such a profile is very useful in both health assessment and design of ergonomic products (Roberts, 1996).

How a machine 'see things' is governed by *computer vision* technology. The term 'computer', incidentally, refers to any computation machinery here, from simple consumer electronics to sophisticated high precision medical image scanners. Computer vision is about identifying and extracting information from digital images through learning and object

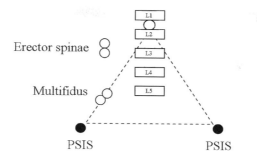

Figure 7.2 Reference points with reference to anterior superior iliac spines (ASIS) and posterior superior iliac spines (PSIS)

recognition. To begin, how can a computer identify a human body from the background? The human body does have some kind of generic shapes, but they can differ significantly, as shown by the three examples in Figure 7.4. All three seem very obvious to us that they are images of a person standing. To us, we can easily tell that the left and centre features the same person, and the image on the right is the sketched figure of that in Figure 7.1. However, a computer visualizes things very differently. A computer relies on algorithms that extract features, objectives, and any specific activities. The body shape, perceived as a 3-D image to the machine vision algorithm, can be manipulated by pattern recognition and feature extraction mechanisms such as those described by (Ezguerra and Mullick, 1996). Given the number of options available, the transmission efficiency should be thoroughly considered particularly for users who move around making detection even more difficult (Pankaj, 2002). Feature extraction is usually necessary when the image is too large so that any information not related to the

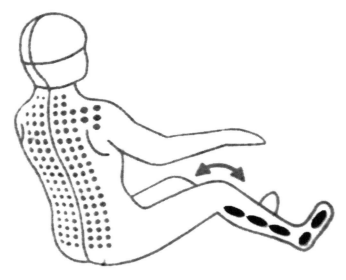

Figure 7.3 Body profile

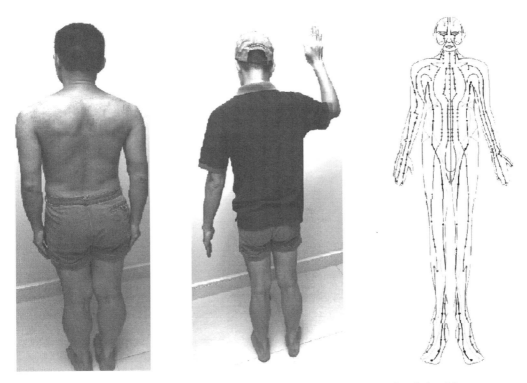

Figure 7.4   Three human pictures not easily recognized by computational algorithms

user's body is extracted and removed, leaving behind only relevant information for analysis. The remaining data contains a description of the body's curvature, which conveys information such as edge direction and shape information (Hoshiai, 2009). The information is then mapped to a generic body contour profile. Apart from imaging methods that reassemble what an eye sees, other methods such as sensors that press against the user so that the distance of various points can be measured against the relative position to a flat plane. This can produce a set of information about the body shape.

So how do these technologies benefit the medical and healthcare industry? Various consumer appliances for general fitness and well-being to be discussed later in this chapter use these to apply certain therapies on specific areas of the user's body. Another major application found throughout the world is removal of excess fatty deposits under the skin for body trimming in cosmetic surgery. Many are willing to spend hundreds or even thousands of (sterling) pounds on removing a few pounds of body weight. Ultrasonic and laser liposuction have been used for shaping body contour through tightening skin surface and removing fat. The former injects a saline fat-loosening fluid into the area concerned and the ultrasonic wave provides energy for melting the body fat. Ultrasound is also used in *vaser liposuction* as a non-invasive surgery with local anesthesia which numbs the area concerned and flushes out the fat. Laser liposuction inserts a tiny tube into the area where the excess fatty deposits are located and eradicated them with the laser beam. Due to its highly focused beam surrounding tissues are not affected. Laser is mainly used for areas not as accessible such as the back of an arm or the inner thighs. As with most types of surgeries, the patient's medical history should be retrieved prior to operation since patients with ailments such as high blood pressure, diabetes or cardiopathy may risk complications in addition to pulmonary fat embolism, perforation of viscuous tissue, edema or swelling.

The acupoint detection process is even more complicated than computing the body contour as acupoints are relatively small and they can sometimes be located very close to each other. Apart from mapping graphically based on a reference chart, (Liu, 2007) described a method using electric characteristics of an acupoint's various anatomic layers for detection. Having briefly discussed the technology related to identifying body contour and to locating acupoints, we shall look at using acupressure for providing temporary relief for an emergency with reference to (Yeung, 2000).

7.1.3 Temporary On-Scene Relief Treatment Support

Acupoints have different healing properties. (Bock, 2009) has reported that some acupoints provide relief by warming up and stretching routines to help the body prepare for training. Acupoints also differ in perceived effectiveness, i.e. some can be readily felt with swift response upon application of force, while others may yield more long term but slower effects. More than one acupoint may be 'linked' to an organ and these points for the same organs may not necessarily be located in close proximity. As such, any effort in providing temporary relief using acupressure requires thorough knowledge of properties for many acupoints.

As an example, we look at a case study on treating sea sickness, a type of motion sickness, as a normal response to movement to the body. Different individuals respond to perceived or actual movement differently (Riccio, 1991). In some cases, the inner ear may sense rolling motions that the eyes do not see. Sometimes symptoms may return momentarily even after the motion stops. Motion sickness can cause anxiety, dizziness, nausea or vomiting. Although medication can control motion sickness, there are circumstances which someone may not have been prepared for, such as unusually rough seas in a normally calm area. Medications for motion sickness such as scopolamine and promethazine are not always suitable due to side effects that may risk blurred vision, drowsiness and impaired judgement. Although biofeedback and cognitive behavioural therapy are reported to be effective for managing motion sickness (Dobie, 1994), the former utilizes instruments recording skin temperature and changes in muscle tension and the latter relies on exposure to a provocative stimulus usually on a specially designed chair; these are not readily available tools that one can easily access when the need arises. Acupuncture at the P6 or Neiguan point to relieve motion sickness is reported to be effective (Stern, 2001). In addition, (Barsoum, 1990) has reported unplanned antiemetic injections could be reduced by acupressure.

Although (Miller, 2004) did not find any concrete scientific evidence of linkage between acupressure and motion sickness, acupressure is still widely used by yachtsmen (Shupak, 2006). This is an area where telemedicine can be very helpful in providing temporary relief treatment support. Off-shore support can only be provided by wireless links as this is the only practical method of obtaining anything instantaneously. Having acquired the underlying

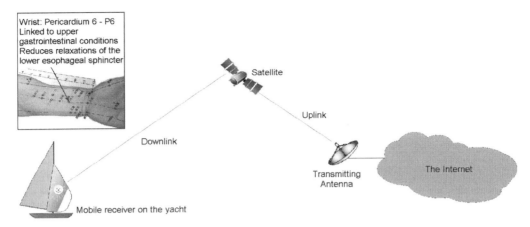

Figure 7.5   Telemedicine for off-shore relief support

concepts of telemedicine in the earlier parts of this book, readers should be able to sketch the support block diagram shown in Figure 7.5. Here, we can see that treatment information can be delivered onto the yacht through wireless communications. In this particular example we use a satellite link to do the job. So, why should we spend time on discussing this example as the system is so simple? To answer this let us conclude this sub-section by going deeper into what is happening here when someone wants to provide off-shore acupressure treatment.

Remember, in sub-section 7.1.1 we mentioned that acupoints serving the same purpose have varying healing properties; some provide swift response while others may not have any immediate response felt. Any attempt to memorize the characteristics of a set of acupoints would be impractical, as this would be analogous to learning an entire pocket dictionary by heart. In practice we need some kind of database that is accessible onboard. Essentially, a database containing information about the acupoints should be searchable and accessible across the Internet. The information received includes location of the appropriate acupoint, where it is linked and what it relieves, as shown in Figure 7.5. Such information is given to yachtsmen who may not have any practical knowledge of acupressure, but due to its popularity amongst the boating community the first time they attempt practicing it may well be when a genuine need arises. Information may therefore be delivered in the form of interactive tutorial which shows how and where pressure is exerted based on needs. However, the amount of data shown, after compression, is likely to be tens of megabytes (MB) for this kind of illustration. This may not be practical with satellite communication due to its inherent delay.

7.1.4   Herbal Medicine

Telemedicine on herbal medicine is arguably one of the most important applications that IT supports. Although it is not necessarily true that telemedicine is widely used in herbal medicine, its origin certainly forms the basis of remote support for modern medical science as we described in section 1.1 over some 5 000 years of ancient telemedicine history. It started when treatment solution was delivered from its source to destinations.

An herb is a plant, or part of, that possesses medicinal, aromatic or savoury qualities. Many popular medications used today were developed from ancient healing traditions that provided a cure using specific plants (Weiss, 2000). The healing components of a given herb are extracted and analyzed for pharmaceutical exploitation. One good example that directly resulted in the 5 000 years old respiratory treatment written in 2735 BC by the Chinese emperor Shen Nong would be the bronchodilator *ephedrine*, extracted from the plant ephedra as a decongestant. This evolves to pseudoephedrine, as its synthetic form, now applied in many allergy, sinus, and cold-relief medications mass produced by the pharmaceutical industry. The link between herbs and modern pharmacology is so close that as many as 40% of the prescription medicines dispensed in the US contain at least one active ingredient derived from herbs (Wilson, 2001). The vast majority of these drugs is either made from plant extracts or is synthesized to fabricate a natural plant compound.

The first formal documentation of herbal medicine is probably the Translation of the Dispensatory entitled '*A Physical Directory*' by Nicholas Culpeper, circa 1649. The use of herbs for treatment became popular throughout Europe and beyond since its documentation. In the modern world, telemedicine allows information about herbal materials to be gathered from remote forests for analysis of substances that act upon the human body and study for contraindications and any possible side effects. The compositions of plants contain many substances including vitamins and minerals; one important aspect of technology is to ensure that the amount of intake of any component does not exceed a toxic level that may result in health damage rather than healing. According to the Natural Resources Conservation Service of the US Department of Agriculture (USDA, NRCS, 2009), as many as hundreds of thousands of plant species exist. Identification and subsequent isolation of active ingredients would entail thorough studies of individual plants. The vast number of plant species mean only a small number have been studied for their healing properties. Further to active ingredients, synergistic interactions between different components within the plant also need to be studied in order to grasp a comprehensive understanding of its medical value. Botanical study will continue to play a significant role in pharmaceutical research in the foreseeable future. Likewise, related telemedicine technologies will be an important part in supporting such research work. This is due to the fact that herbal medicine does not have the same level of acceptance in all countries (Chitturi, 2000). Cross border study facilitated by information exchange would certainly accelerate the lengthy process of botanical study.

7.2   Interactive Gaming for Healthcare

Over the past couple of decades, children and adults alike have been addicted to video games. A good 'workout' on the video game console is likely to keep a user there for longer than on a treadmill. Many video games make users concentrate on playing them for an extended period of time. This prolonged continual exposure to computer or TV screen will very likely degrade eyesight and increase the risk of glaucoma, an insidious disease that affects particularly those who are short-sighted. The impact of glaucoma may result in loss of peripheral vision over time. In addition, (Kasraee, 2009) also reported a confirmed case of ideopathic eccrine hidradenitis, a skin disorder that affects the game player's hands. The gripping of game controller may well have contributed to sweating that causes swollen sweat glands in the palm. The physical symptoms add to the known psychological effects of video game addiction that together make video gaming famously unhealthy. Further, the user's posture may also contribute towards acute tendonitis and back pain. Having discussed all the health problems related to video games, we may ask ourselves why bother talking about telemedicine related technologies for gaming since they appear to contradict each other. The first obvious answer would be exercising common sense by monitoring playing time and take regular breaks. Of course, there must be some incentives to warrant a thorough discussion on how video games can also be made to improve fitness and well-being.

There is a growing trend for children to become overweight, partially contributed by a combination of lack of exercise and unbalanced diet. Another prevalent health hazard is that schoolbags are too heavy for many school children and uneven weight distribution within a schoolbag can cause spinal injury over time. While the links between obesity in children, lack of physical activity and video games have been blamed as something unhealthy for decades, properly designed games with fitness in mind can provide more than thumb exercises one usually gets with playing video games. Combining interactive gaming and remote health monitoring technology can help maintain optimal health and reduce consequences associated with being overweight and obesity, which is becoming an increasing problem in many metropolitan cities; children are far less likely to develop medical conditions that put further pressure on the public healthcare system. Unlike conventional computer games where user control merely involves hand movement, these games have specific design considerations such that movement of the entire body will stimulate physical exercise.

7.2.1   Games and Physical Exercise

Video games, involving physical interaction between human and computer, usually engage a controller such as a joystick that has been widely used for decades, on which a user presses several buttons in rapid succession. For other different control methods, there are workout activities, strength training with a balance board, and so on (Robertson, 2008). The concept of fitness gaming is not something new, several major Japanese video game manufacturers have already launched a number of workout games over several years (Brandt, 2004). Some workout games have features that record calories burned and the distance one would have covered to get the equivalent results during a gaming session. In a small room, one can play a range of simulated sporting activities by standing on a mat with an array of sensors. The player's movement will be relayed to the TV screen via a game console. For example, skiing can be simulated by detecting body lean as the movement and translates to the direction of which the player is leaning which will then be relayed and displayed on the TV screen as illustrated in Figure 7.6. The mat is connected to the game console via a wireless link so that there is no risk of tripping over tangled wires while playing.

7.2.2   Monitoring and Optimizing Children's Health

Although some game consoles feature automatic tracking of parameters related to user fitness such as BMI, measuring size of chest, biceps, waist, and thighs, etc. Technology can do far more than this. A comprehensive health monitoring system can be built on the paradigm shown in Figure 7.7 that consists of three separate modules.

Figure 7.6 Virtual skiing video game

1. Gaming Console:

    Basic features include computing the amount of physical exercise completed during school, such as inclusion of scheduled activities sessions. These games will also be categorized into individual and group activities so that children can play alone or together either at their own homes physically separated or gathered together. Unlike most off-the-shelf computer games where user control is accomplished through joysticks involving movement of very

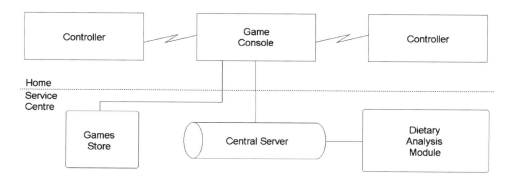

Figure 7.7 Gaming system for physical exercise

limited parts of the body, these games will be controlled by a small body area network (BAN) installed in various positions of the user's limbs that track their movements, thereby providing their scoring and feedback on amount of exercise completed. This module mainly consists of three major parts:

The games (software), supported by appropriate controllers, have to be made suitable for different ages with varying intensities of physical demands. Also, graphic design must be tailored for the intended age of users. The games are played by body movement when users undertake physical activities. Artificial Intelligence (AI) is part of the underlying technology that determines the most suitable game for an individual user, selected based on user parameters such as age, BMI, time since the previous meal, whether a physical exercise lesson at school has been scheduled on the day. While games are generally developed for children according to age groups, an option should be provided for boys and girls since children of different genders may prefer games of different themes.

2. Controllers:

Game control for fitness and workout involves complex detection of movement with various parts of the body in order to serve the primary purpose of promoting physical exercise. A BAN with sensors will be set up for movement tracking. Haptic sensing for touch will also be used for the control console and menu navigation. A wireless receiver captures the data from the BAN. In addition to body sensor networks, video imaging technique that analyzes the user's movement based on object extraction and frame analysis can also be used on a commercial scale. This is not practical for consumer applications because the camera installed and complication in evaluation of amount of force exerted by the limbs during the exercise may be very costly.

3. Dietary Analysis Module:

Recommended diet provides optimal levels of intake of essential nutrients. This can be accomplished on the basis of information enlisted by the Food and Nutrition Board of the National Academy of Sciences to be adequate to meet the nutrient needs of a child of a certain age to ensure consumption of adequate amounts of essential nutrients. References such as the Food Guide Pyramid (FGP) can be as a tool for healthy food choices. Key guidelines include not exceeding 30% of total energy intake from fat and getting less than 10% from saturated fats. The FGP for young children (two to six years old) identifies recommended portions of foods from grains (six servings), vegetables (three servings), fruit (two servings), milk (two servings), and meat (two servings), as well as recommending limiting the intake of fats and sweets. The nutrient needs of teens can be determined using the FGP for adults. These will be used as references when calculating the optimal dietary recommendation for a user. Users will be asked to enter what has been provided by school. Without any knowledge of quantities of each type of food, estimation needs to be computed from a list of pre-programmed combinations so as to derive an estimate of the intake from a school meal. This module will be a computer programme that takes user input on the contents of a school meal, estimates the amount of each component and therefore the nutrition composition, thereby produces a list of recommended food for snacks and dinner for optimal health.

4. Central Server:

In addition to storing all necessary information supporting the above modules, this module accommodates a database serving individual users through the internet. It provides main administrative and support functions such as information update, games maintenance, and data protection.

7.2.3   Wireless Control Technology

Wireless communication plays a vital role in fitness gaming since users cannot be tangled with wires as they exercise. Controllers may include mats and a variety of handheld controllers. Control can be accomplished by a conventional handheld unit such as those used in Figure 7.6 attached to an array of sensor network. All these have two fundamental attributes in common: (1) they are all battery powered with an onboard power source, i.e. each unit or sensor is self-contained; (2) each has its own wireless transmitter, usually implemented in an SoC (System-on-Chip) configuration, i.e. control data is transmitted to the console through a wireless link by a single IC (integrated circuit) chip that contains components such as transceiver, data buffer, and filters.

Although Bluetooth is used in most commercially available wireless gaming devices (refer to section 2.2 for details), the Xbox 360® does not (as of July 2010). Readers who have read Chapter 2 should know by now why Bluetooth is a good choice; its properties such as low power and eliminating the need for line-of-sight certainly make it suitable for game controllers. However, there are certain requirements and limitations with the Bluetooth standard. The video games industry is so huge that over 20 million game consoles are sold per year (Rosenberg, 2009). This enormous sales figure, shared by three major market players, can well support the development of proprietary standards for data communication that can be customized exclusively for specific applications with regulatory compliance being just about the only constraint. This leads to a challenge of balancing between transmission power and data throughput. The transmission power should be minimized to prolong battery life while maintaining adequate data throughput for the data collected from movement. As these controllers are battery-operated, the system must be developed to eliminate the risk of sudden loss of control due to battery exhaustion. Rechargeable batteries for consumer electronics are primarily made using Nickel Cadmium (NiCd), Nickel Metal Hydride (NiMH) or Lithium Ion (Li-Ion) cells.

Before we conclude this section, we take a quick look at different types of batteries that can be used in these wireless controllers. Compared to the oldest NiCd type found in the 1970s, NiMH has about twice the energy density and it does not have any *memory effect*. Memory effect is a term used to refer to a battery that only retains a proportion of its maximum capacity after repeatedly being only partially discharged before recharging. In addition to weight and capacity advantages, NiMH batteries are also more environmentally friendly than the NiCd as they do not contain heavy metals and the chemicals used are less toxic. Li-Ion, becoming increasingly popular over recent years and found in most of the latest appliances, produces the same energy as NiMH batteries but weighs one-third less. Physical movement of the controller may be a charging method to recharge the battery while in use. However, each type requires a different charging pattern to be properly recharged such that some may require continuous electric current while others may be bursty with pulses at a certain time interval. The life of a rechargeable battery operating under normal conditions is generally less than 1000 charge-discharge cycles, such that the user will experience a decline in the running time of the battery due to aging. Mechanisms that provide battery health monitoring would be helpful for determining the health status of the battery (Gu, 2009). Emerging technology utilizing Ruthenium Oxide may soon become available which has many advantages over the three major cells discussed above. As of January 2010 this type of high capacity thin-film battery is only manufactured for special purpose apparatus by Flexel® in areas such as wireless sensor networks and implantable devices. Its properties, well suited for wireless controllers, may become a durable solution in the near future.

7.3 Consumer Electronics in Healthcare

An earlier report by (PR Newswire, 2007) suggested that the consumer electronics market for therapeutic and well-being devices and services will be worth US$4 billion plus another two billion for the online well-being market by 2010 with a yearly growth rate of 20%. This report includs many types of devices ranging from blood pressure meters to wireless healthcare gaming with examples such as the Nintendo® 'Brain Age' software for memory retention and mental fitness. As emergence of new generation wireless devices provide capabilities for supporting more healthcare services, market growth will certainly continue over the years to come. Consumer healthcare technology is not restricted to linking between manufacturers and consumers, entities such as government agencies and insurance companies also have direct interests as the former promote general health awareness so as to reduce avoidable hospitalization and other healthcare services; the latter will certainly benefit from fewer insurance claims by addressing fitness and well-being.

Telecare products business may be more challenging under which many consumer healthcare products will be wireless enabled in the foreseeable future. For example, a wide range of healthcare applications are available for Java-enabled mobile phones connected to a 3G network. The availability of existing network infrastructure and wireless connection, especially in rural areas, would certainly affect sales growth. Also, these services rely on co-operation from vendors such as Internet service providers (ISP) who may not be willing to guarantee connections for data transmission and availability.

7.3.1 Assortment of Consumer Appliances

A very wide range of consumer electronics products are readily available from stores all around us. These range from a small electric toothbrush of £2 (US$ 3) to a giant luxury massage chair with a £5 000 (US$8 000) price tag. Literally everything that covers the entire body from head to toe. The selection is so vast that even appliances like hair dryers and shavers are claimed to be healthcare products. Loosely speaking, other devices readily available from a local appliance store like blood pressure monitors, digital thermometers, massaging apparatuses, and grooming devices are all healthcare related. Whether these devices are really related to healthcare is not up for discussion within our scope. With so many healthcare products around we are unable to cover every single category in one book chapter. We have no intention of providing extensive coverage for each type of product; instead, we shall primarily focus on the underlying technologies. There are some common attributes for consumer electronics products. First, they are mass produced, meaning that minimizing the manufacturing cost of each unit while maintaining maximum reliability is vitally important, parts selection therefore plays a vital role in this aspect (Pecht, 2005). Unlike most consumer electronics appliances, healthcare devices have more direct physical contact with the user. The inherent safety risk would therefore be comparably higher. In our case study, we look at a massage chair with product specifications listed in Table 7.1. What does it have for us here? First, its electrical current can be over 1 A, it must be mains powered. Prolonged standby may also pose fire risks as in numerous cases of TV sets all over the world, such as that reported in (BBC News, 2007). Another potential fire hazard is the massaging rating of 30 minutes continuous operation, which is in fact very common among these products. The unit should be left to cool down after

Table 7.1 Product specifications of a massage chair

Operating Voltage:	110~120V / 50–60 Hz
Power Consumption:	120 Watts to 300 Watts / 5 Watts Standby
Massaging Rated:	30 Minutes (Continuous)
Recline Angle:	115~175 Degrees
Massage Stroke Length:	30 inches / 76 cm
Dimensions:	W33' × D45' × H49' (Upright) / 83 cm × 114 cm × 124 cm /W33' × D75' × H30' (Recline) / 83 cm × 190 cm × 76 cm
Net Weight:	190 lbs / 86 kg
Material:	Fire Retardant PVC Leather

half an hour to avoid overheating. Reclining must be clear of all physical obstacles that can result in motor damage. The next issue to consider is the bulkiness of the unit; ergonomical design for handling should be exercised to minimize the risk of back injury with such large products. Finally, the material used can be damaged by a sharp object; for example, a user may sit on the chair with metal objects in the trouser pockets that cut through the upholstery. Any damage to the device may expose hazardous parts that could lead to physical injury or even electrocution. Also, many devices are made for operation in the bathroom which may also increase the risk of electric shock. So, running down this seemingly short list reminds us of many possible risks that need to be carefully considered during the product design stage. Although many small devices are battery operated so that electric shock is not an issue, there are many other risk factors that must be considered as we shall discuss next.

7.3.2 Safety and Design Considerations

A baby monitor is probably the best product to use in our case study since it is often the very first wireless communication device that a person uses. While the baby does not really know what benefits it brings, the parents can move around the home with the assurance that every activity of the baby will be detected. It is a device that brings the baby and parents close together.

Concern over liability and its litigation is always an issue with healthcare related products as they are often required to comply with different sets of regulations governing the marketing and sales of such products imposed by different authorities. This is an important challenge that many healthcare device manufacturers face because some countries even have different state and provincial requirements. Many manufacturers may seek local partnership prior to importation for overcoming these requirements.

In many metropolitan cities, people are exposed to pollutants, toxins and electromagnetic fields on a regular basis. We may not be able to do a lot to contain environmental pollutions but attempts to reduce electromagnetic emission from electronic devices can reduce the effect of electromagnetic fields radiating from electronic appliances. Power control and proper device shielding would be effective ways to reduce such impact.

EMI is certainly an important issue both in terms of operational reliability and radiation safety to the baby's healthy growth. Of course, no wireless device can be made EMI-free, laboratory testing to determine the design and operating characteristics of the device should be

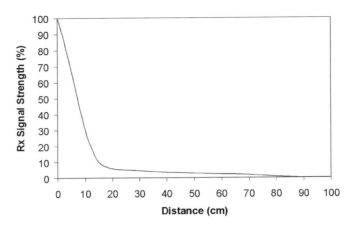

Figure 7.8   Field strength versus distance

carried out. In this regard, imposing a minimum separating distance between the baby and the device can mitigate much of the EMI risk. Active power control that minimizes transmission power output can limit the amount of time that the transmitter is active thereby reducing exposure to radiation by the user. The maximum transmission power should be suppressed within the (FDA, 2009) limit. The proximity also plays an important role in controlling EMI exposure. Figure 7.8 shows the field strength relative to distance. A substantial amount of energy is lost within the first 10 to 20 cm and drops to only 5% at a distance of about 20 cm from the transmitter. Therefore, placing the transmitter 20 cm or 8 inches away from the baby will drastically reduce the radiation associated with EMI since the field strength emission decreases rapidly as distance increases. In sub-section 2.1.4, we discussed the concept of electromagnetic compatibility (EMC) as opposed to EMI, meaning that the device needs to be compatible with the surrounding EM environment so that it can still operate reliably when subject to interference, in addition to satisfying the emission limits of EMI that may affect the operation of other nearby devices. EMI *shielding* therefore becomes necessary. It is also worth noting that older equipment can degrade over time and become more susceptible to EMI.

A baby's skin is very delicate and sensitive. The choice of material and the housing must be very carefully designed to ensure optimal ergonomics and safety. Battery leakage is also a problem that can lead to serious consequences as toxic chemicals can reach the baby. To avoid such risk, the battery compartment should be properly sealed in such a way that toxic chemical will be contained within the monitor in the event of a battery leakage.

7.3.3   *Marketing Myths, What Something Claims to Achieve*

The idea about myths, perspectives, and inversions of marketing a product is perhaps best described with the following characteristics (Vargo, 2004):

- Intangibility:lacking the palpable or tactile quality of goods.
- Heterogeneity:the relative inability to standardize the output of services in comparison to goods.

- Inseparability of production and consumption:the simultaneous nature of service production and consumption compared with the sequential nature of production, purchase, and consumption that characterizes physical products.
- Perishability:the relative inability to inventory services as compared to goods.

Marketing is often used as a tool to deal with these (Zeithaml, 2000). What is said may or may not be true. The effectiveness of a given product may only be tested in a laboratory. Unfortunately for the consumer, this information is usually withheld, opening the door for manufacturers to exaggerate what their products can do. As with all consumer products, the fine print in the warranty will often contradict what the marketing people say about their products. This is particularly the case in a competitive market like consumer healthcare products where many manufacturers offer very similar products. Sometimes marketing materials can be deceptive. Take, for example, this quote direct from the product box of a massage chair: '. . . we've earned a sound reputation for producing durable, reliable, stable, industrial-grade healthcare products over three decades'. Now compare that statement to what is specifically excluded in the warranty terms and conditions:

- Sample of what is NOT covered
- Wear and tear from moving parts
- Commercial and industrial use

So, is it made to be stable and reliable? Is it meant to be 'industrial-grade'? Marketing statements are often misleading and contain no more than superficial gimmicks. Claims are sometimes made based on some studies. For example, certain university studied . . . and confirmed . . ., these studies may be conducted in a well-controlled environment to support a claim by an expert in the field. Playing with psychology is often an effective marketing deception (Boush, 2009).

7.4 Telehealth in General Healthcare and Fitness

Medical technology is not always used for assistive remedy and it is also extremely important when providing a range of solutions for maintaining optimal health. There are many ways that we can keep ourselves healthy with technology, such as dietary monitoring and various massaging devices that apply reflexology to alleviate stress and tension. Further, there are exercise therapies such as yoga and martial arts that enhance circulation and flexibility while easing chronic pain. After all, a healthy lifestyle is about nutrition, exercise, and stress relief. Throughout this chapter, we have discussed how technology has been developed to help us maintain a healthy lifestyle with all these. To conclude the chapter, we shall look at ways telehealth and related technologies can be used to assist us with optimizing our health while exercising.

7.4.1 Technology Assisted Exercise

Being physically active and maintaining a healthy lifestyle benefits people of all ages. Moderate physical activity in our daily lives will certainly keep us in optimal shape and help us worry

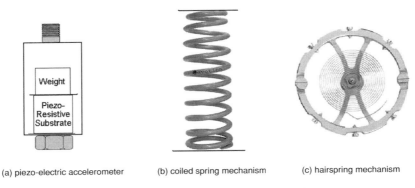

(a) piezo-electric accelerometer   (b) coiled spring mechanism   (c) hairspring mechanism

Figure 7.9   Fundamental components of a pedometer

less about illness or any health problems. Not all physical exercise needs to be taken in a gymnasium because something like a half hour walk can also keep us physically active. Earlier work by Tuomilehto (2001) has shown that physical exercise can reduce the risk of developing diabetes for those at high risk. A wide range of exercise can be taken that fits an individual's schedule. From walking and jogging to swimming and ball games, there are so many technology has to offer.

Technology can ensure exercise is taken at an appropriate and comfortable pace. For example, simple measurement of the heart and respiratory rates can prevent any difficulty in breathing or fainting during or after exercise. For many whom simply take a short walk after dinner, there is a step counter, also known as a pedometer, which counts your steps, determines the distance you have covered and your calories burnt while walking. A pedometer mainly works by sensing body motion and counting the number of footsteps. All pedometers count steps, although they may have different counting methods. These can be piezo-electric accelerometers, a coiled spring mechanism, or a hairspring mechanism. As illustrated in Figure 7.9, all these operate by compression and subsequent expansion. Each cycle translates to one step count. With knowledge of the user's usual stride length, the count can be multiplied by the nominal stride length to obtain the distance covered. The simplest pedometers only count your steps and display steps and distance. This can even be implemented on some mobile phones as a built-in feature that works by relaying data transmitted from pedometer sensor. When designing a pedometer, it is important to ensure that the count reset button is not easily depressed unintentionally and sufficient memory is available to store the number of steps for a specified number of days. Some even obtain geographical information from GPS for tracking continuous speed and distance therefore speedometers and odometers just as those found on vehicle dashboards are also available. The user can download and overlay the workout to a map and obtain information about the elevation and gradient of the hill that they climb. There is, however, a potential problem with tracking while walking under trees or tall buildings that may cause the wireless communication link from the GPS to be temporarily disrupted.

GPS technology can also assist cyclists. A GPS unit can replace a conventional cycle computer to provide features such as route mapping, recording heart rate, and pulse data that can be downloaded to a computer for fitness analysis. The electronics can do just what the control panel of a gym exercise bike can offer.

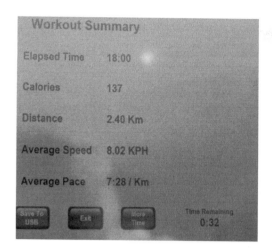

Figure 7.10   Workout summary

7.4.2   In the Gymnasium

There are many choices of fitness equipment in a typical gymnasium. A gymnasium usually features treadmills, exercise bikes (include elliptical cross trainers), rowing machines, boxing, weights and bars. While how they operate and what potential benefit this equipment brings is not within the scope of this text, we intend to explore the technologies that support these devices.

The treadmill, an example, is perhaps the most likely to break down, so reliability becomes a primary issue. The motor is the heart of a treadmill, the major cause of a treadmill motor failure is high walking belt friction created by a lack of lubricant. The cost of replacing a motor is usually around £120 (US$ 200), and preventive maintenance can certainly prolong the motor's life significantly. Condition based prognostic management can provide a solution to service the motor prior to an anticipated failure. We shall discuss this in more detail later in section 9.1. Apart from keeping the treadmill up and running, this seemingly simple to operate equipment is fully packed with technology. Apart from controlling the speed of the walking belt, which also can cause a reliability issue since this is the part that wears out most quickly due to persistent rubbing between the shoe sole and its surface, a treadmill has a lot of features to offer. Upon completion of a workout, a set of statistics can be collected as in Figure 7.10, which show the duration, amount of calories burnt computed from the effort of the workout, distance covered and elapsed time. This information can then deduce the average speed and pace. Note also that there is an option to 'Save To USB'. The data can be stored for analysis to keep track of the user's fitness. It is also possible to automatically download the data via a wireless link.

One common feature found in most equipment is the handgrip-mounted heart-rate monitor shown previously in Figure 4.4 that helps the users to calculate the calories they are burning during the workout and see whether they are working hard enough or too hard. Other common features include simulated workout profile such as the cycling path profile shown in Figure 7.11.

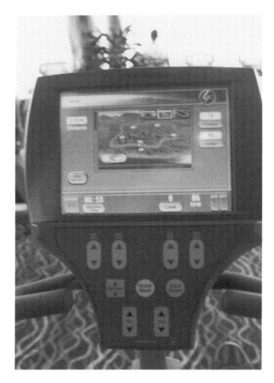

Figure 7.11   Simulated cycling path profile

A number of devices such as heart rate and blood pressure monitors can also be wirelessly linked to collect different body signs for health assessment.

Having covered what we commonly use in the gymnasium, let us conclude by taking a brief look at telehealth technology used in professional sports training. An array of small sensors can be attached to the user to collect real-time data about a sports training session, the performance can be analyzed and recorded by a computer. There are many types of sensors to meet different needs, for example, boxing requires an accelerometer for gait and posture analysis; pressure sensors to track where the user has been hit by what amount of force. Such biomechanical measurements can quantify the functional performance during training. This can also be accomplished by video motion tracking. Another area is surface electromyography (EMG) analysis that measures muscle contraction that initiates limb movement. This quantifies the muscle activity and fatigue.

7.4.3   Continual Health Assessment

Telehealth is proven to be extremely helpful for those patients who need the most frequent contact, where as many as 7.6 million people in the USA alone are reported to be receiving home care because of acute illness, long-term health conditions, permanent disability, or terminal illness (NAHC, 2008). Upon discharge from the hospital to their homes, patients with chronic disease often require close surveillance throughout the rehabilitation process. Telehealth enables medical professionals to monitor the patient continuously, make real-time identifications and interventions in the care of their patients. Although telehealth is extremely important for those with special needs, its benefits also extend to normal healthy people. In addition to what we have discussed on how telehealth and related technologies assist us with daily exercise, technology can also help us maintain optimal health.

So far we have talked about using various meters to check body signs, there is more we can do than analyze these figures to ensure that we are maintaining our health optimally. For example, we can keep track of what we eat and balance the nutrition intake. We can record the amount of food eaten and compare that against how much exercise we have undertaken on the day, to ensure that we do not have fat build up. Under the assumption that the food packaging provides true nutrition information about its contents, we can use such information to log the amount of food we have taken with a breakdown of individual components such as salt and carbohydrates. An optimal meal can also be determined with reference to the Food Guide Pyramid (FGP). A software-based FGP Automated Analysis System such as that proposed in (Muthukannan, 1995) can be installed on a mobile phone for continual analysis. This information may be nice to know for a healthy person but can be extremely helpful to those who need to control their diets for a variety of reasons. For example, this helps a patient with kidney problems to ensure no excess sodium is taken.

We have seen many healthcare applications that do not utilize traditional medical science and technology for both healing and maintaining general health. Although medicine will certainly continue to be the mainstream cure for all of us, there are many alternative ways that technology can be used for healthcare as we have seen. The popularity of this diverse range of CAM makes us look into how telemedicine and related technologies can be developed for better health.

References

Abreo, V. (2009), Acupressure Daily Exercise, *BellaOnline Alternative Medicine Site*, http://www.bellaonline.com/articles/art60875.asp

Bansoum, G., Perry, E. P., and Fraser, I. A. (1990), Postoperative nausea is relieved by acupressure, *Journal of the Royal Society of Medicine*, 83(2):86–89.

Barrett, S. (1993), Complementary self-care strategies for healthy aging, *Journal of the American Society on Aging*, 17(3):49–53.

BBC News (2007), TV on standby 'caused explosion', 20 March 2007.

Bock, D. (2009), Acupressure points for stretching & releasing tension, *EC Martial Arts Blog*, 21 February.

Brandt, A. (2004), The game room: gaming for fitness, *PC World*, 7 February.

Bratman. S. (1997), *The Alternative Medicine Sourcebook*. Lowell House, ISBN: 1565656261.

Boush, D. M., Friestad, M. and Wright, P. (2009), *Deception in the Marketplace*, Routledge Academic, ISBN: 9780805860870.

Chitturi, S. and Farrell, G. C. (2000), Herbal hepatotoxicity: An expanding but poorly defined problem, *Journal of Gastroenterology and Hepatology*, 15(10):1093–1099.

Dobie, T. G. and May, J. G. (1994), Cognitive-behavioral management of motion sickness, *Aviation, space, and Environmental Medicine*, 65(10 Pt 2):C1–2.

Ezquerra, N. and Mullick, R. (1996), An approach to 3D pose determination, *ACM Transactions on Graphics*, 15(2):99–120.

FDA (2009), Medical Devices and EMI: the FDA Perspective. http://www.fda.gov/Radiation-EmittingProducts/RadiationSafety/ElectromagneticCompatibilityEMC/ucm116624.htm

Fong, B., Yeung, E. H. K., Chow, D. H. K., and Pecht, M. G. (2010), Electromyographic Study of Erector Spinae and Multifidus due to Use of Elbow Crutch, - *submitted for publication*.

Food Guide Pyramid: http://kidshealth.org/kid/stay_healthy/food/pyramid.html

Gu, J. and Pecht, M. (2009). Health assessment and prognostics of electronic products: an alternative to traditional reliability prediction methods, *ElectronicsCooling Online*, http://electronics-cooling.com/html/2009_may_a1.php

Horstmanshoff, H. F. J., Stol, M. and Van Tilburg, C. R. (2004), *Magic and Rationality in Ancient Near Eastern and Graeco-Roman Medicine*, Brill Publishers, ISBN: 9004136665.

Hoshiai, K., Fujie, S. and Kobayashi, T. (2009), Upper-body contour extraction using face and body shape variance information, *Lecture Notes In Computer Science*, Springer-Verlag Berlin, Heidelberg, 5414:862–873, ISBN: 9783540929567.

Kasraee, B., Masouye, I., and Piguet, V. (2009), PlayStation palmar hidradenitis, *British Journal of Dermatology*, 160(4):892–894.

Liu, T. Y., Yang, H. Y., Kuai, L. and Gao, M. (2007), Problems in traditional acupoint electric characteristic detection and conception of new acupoint detection method, *Zhongguo Zhen Jiu*, 27(1):23–25, http://www.ncbi.nlm.nih.gov/pubmed/17378198

Lu, G. D. and Needham, J. (1980). *Celestial Lancets: A History and Rationale of Acupuncture and Moxa*, Routledge-Curzon, New York, ISBN: 0700714588.

Maa, S. H. et al. (2003), Effect of acupuncture or acupressure on quality of life of patients with chronic obstructive asthma: a pilot study, *Journal of Alternative and Complementary Medicine*, 9(5):659–670.

Marshall, V. W. and Altpeter, M. (2005), Cultivating social work leadership in health promotion and aging: strategies for active aging interventions, *Health and Social Work*, 30(2):135–144.

Miller, K. E. and Muth, E. R. (2004) Efficacy of acupressure and acustimulation bands for the prevention of motion sickness, *Aviation, space, and Environmental Medicine*, 75(1):227–234.

Muthukannan, J. (1995), The food guide pyramid approach to an automated analysis of foods, *Journal of the American Dietetic Association*, 95(9):A49.

NAHC (2008), *Basic Statistics About Home Care*, National Association for Home Care & Hospice, http://www.nahc.org/facts/08HC_Stats.pdf

Nakao, M. et al. (1997), Clinical effects of blood pressure biofeedback treatment on hypertension by auto-shaping, *Psychosomatic Medicine*, 59:331–338.

Nunn, J. F. (2002), *Ancient Egyptian Medicine*, Red River Books, ISBN: 0806135042.

Nutrition Business Journal (2009), Alternative Medicine Market Research Reports, PentonMedia, Inc.

Pankaj, K. et al. (2002), Algorithmic issues in modeling motion, *ACM Computing Surveys*, 34(4):550–572, December.

Pecht, M. (2005), *Parts Selection and Management*, John Wiley & Sons (USA), ISBN: 9780471476054.

PR Newswire Europe Ltd. (2007), *eHealth And Consumer Electronics*.

Rawlinson, G. (1956), *The History of Herodotus*; translated from the ancient Greek, Tudor, ASIN: B001E2SV1I.

Riccio, G. E. and Stoffregen, T. A. (1991), An ecological theory of motion sickness and postural instability, *Ecological Psychology*, 3(3):195–240.

Roberts, D. E., Reading, J. E., Daley, J. L., Hodgdon, J. A. and Pozos, R. S. (1996), A Physiological and Biomechanical Evaluation of Commercial Load-Bearing Ensembles for the U.S. Marine Corps, Report No. 96-7 of the Naval Health Research Center, San Diego CA.

Robertson, A. (2008), Can You Really Get Fit with Wii Exercise Games?, *WebMD Feature*, May 30, http://www.medicinenet.com/script/main/art.asp?articlekey=90064

Rosenberg, D. (2009), Nintendo crushes it in 2008 console sales, *Cnet News*, January 16.

Shupak, A. and Gordon, C. R. (2006), Motion sickness: advances in pathogenesis, prediction, prevention, and treatment, *Aviation, Space, and Environmental Medicine*, 77(12):1213–1223.

Stern, R. M., Jokerst, M. D., Muth, E. R., and Hollis, C. (2001), Acupressure relieves the symptoms of motion sickness and reduces abnormal gastric activity, *Alternative Therapies in Health and Medicine*, 7(4):91–94.

Tuomilehto, J. et al. (2001), Prevention of Type 2 diabetes mellitus by changes in lifestyle among subjects with impaired glucose tolerance, *New England Journal of Medicine*, 344(18):1343–1350.

Unschuld, P. U. (2003), *Huang Di nei jing su wen: Nature, Knowledge, Imagery in an Ancient Chinese Medical Text*, University of California Press. ISBN: 0520233220.

USDA, NRCS. (2009), The PLANTS Database (http://plants.usda.gov, 13 December 2009). National Plant Data Center, Baton Rouge, LA 70874-4490 USA.

Veith, I. (1972), *The Yellow Emperor's Classic of Internal Medicine*, University of California Press, ISBN: 0520021584.

Vargo, S. L. and Lusch, R. F. (2004), The four service marketing myths remnants of a goods-based, manufacturing model, *Journal of Service Research*, 6(4):324–335.

Versel, N. (2009), Wireless home health said to grow nearly 15-fold by 2013, *Fierce Mobile Health-care*, 11 August, http://www.fiercemobilehealthcare.com/story/wireless-home-health-said-grow-nearly-15-fold-2013/2009-08-11?utm_medium=nl&utm_source=internal

Weiss, R. F. (2000), *Herbal Medicine*, 2/e, Beaconsfield Publishers Ltd., UK., ISBN: 0865779708.

Wilson E. O. (2001), *The Diversity of Life* (Kindle Edition), Penguin Publishing, ASIN: B002RI9IRY.

Yeung, P. C. T. (2000), Acupressure in First Aid: 18 Points for Treating Common Emergencies, Healthphone.com Inc. Canada, http://www.healthphone.com/consump_english/a_energy_train/first_aid/index.htm

Yuan, R. and Lin, Y. (2000), Traditional Chinese medicine: an approach to scientific proof and clinical validation, *Pharmacology & Therapeutics*, 86(2):191–198.

Zeithaml, V. A. and Bitner, M. J. (2000), *Services Marketing: Integrating Customer Focus Across the Firm*, 2/e, McGraw-Hill Boston, ISBN: 0071169946.

8

Caring for the Community

The primary objective of any healthcare service is to make everyone safer, healthier, and live longer. Prevention often reduces the need for medical treatment. Telemedicine therefore seeks to provide advice and support to reduce the chance of illness or injury. Historically, technology evolves over centuries to provide people with a better future. This is exactly why we are interested in enhancing medical and healthcare technology to people everywhere by providing more efficient and affordable services with ease of access to as many people as possible.

Technology benefits both service providers and patients in many ways involving physicians, healthcare professionals, end users, engineers, equipment manufacturers, authorities, care centres, clinics and hospitals. Healthcare services are no longer limited to certain locations such as clinics and hospitals as communications technology is able to bring many of these services away from clinics and hospitals to users on the move or at home. Caring for the community is about assisting disabled people, taking care of the children and the elderly, healing the sick or injured, and supporting vulnerable individuals.

Over the past seven chapters, we looked at many different types of wireless communication technologies being used in many different applications that cover the entire human body. In this chapter, we shall look at various aspects of healthcare that are used in caring for people under different circumstances.

8.1 Telecare

Through technical advances in telecommunications, many people who require special attention can live alone with the assurance that help is always available and they are taken good care of. Although telecare focuses more on subsequent responses to a situation rather than actively preventing an event from happening, caregivers can easily find out their whereabouts and attend to them whenever need arises. Sometimes, telecare can even save the caregiver's visit because help can simply be provided remotely. Telecare puts the two otherwise contradictory attributes, independence and monitoring, together in a mutually supporting way. Very simply, people can enjoy the freedom of being left alone while knowing that assistance is always there if or when needed.

Telemedicine Technologies: Information Technologies in Medicine and Telehealth Bernard Fong, A.C.M. Fong, and C.K. Li
© 2011 John Wiley & Sons, Ltd

Customization is a key feature of telecare, necessary tools are provided to a user based on individual needs. Telecare can be as simple as an alarm to call for emergency assistance, or a sophisticated system that monitors the user's health condition, assistive network of devices for various routine tasks, an automated personal assistant that reminds the users of different things such as taking prescribed medication and switching off the gas stove after cooking; and the list of what telecare can do goes on. Telecare also includes a communication link that connects the user to a clinician or response centre for alerts, health monitoring of vital signs, and personal advice. As a brief summary, telecare involves using telecommunications technology for health monitoring and to provide on-demand caring support.

The advantages brought to both users and caregivers by telecare are evident. With so many features to offer, telecare involves the use of different technologies put together. We look at the building blocks of telecare in this section by first introducing the term 'telehealth', which is said by the US Health Resources and Services Administration office to improve access to quality healthcare.

8.1.1 Telehealth

Telehealth is widely considered as a sub-set of telecare with the specific purpose of body vital sign monitoring (Li, 2006). It brings technology and advancing clinical practice together to collect patients' information for monitoring with continuous feedback as well as scheduling for appointments. General health assessment is one key feature promoted by telehealth that uses a wide range of devices covering virtually all parts of the human body. Figure 8.1 shows a number of commonly used ones and they can all be connected as a small mobile healthcare centre. Telehealth provides comprehensive coverage for the entire human body, it also enables patients to perform their own tests that automatically update their electronic patient record. This is particularly helpful for patients while waiting for their medical consultation. With

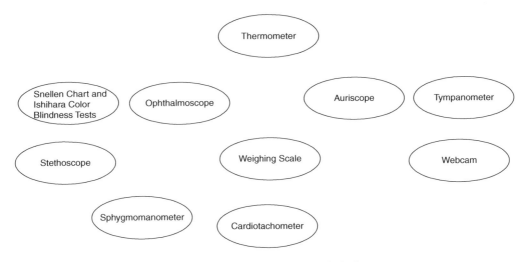

Figure 8.1   Collection of telehealth devices

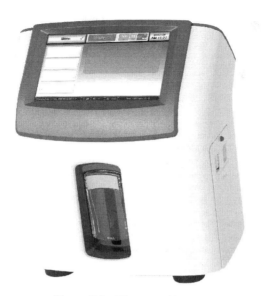

Figure 8.2 Pharmacy kiosk

the range of possibilities shown in Figure 8.1, basic parameters such as body temperature, body mass index (BMI), oxygen saturation, and heart rate can all be swiftly obtained and automatically made available to the doctor when the patient comes.

Another key feature of telehealth is that automated health survey about a patient's current state can be sent back to hospital and also update the electronic patient record. Intake of any pharmacy medicine bought over the counter by the patient without a prescription can also be recorded. The device shown in Figure 8.2 provides a touch screen for conducting medical surveys with a barcode reader that scans the barcode of any medicine that the patient may have taken. By linking to the pharmacy's database, detailed information about the medicine can be known.

8.1.2   Equipment

Due to the diversity of applications offered, there are many different types of equipment for telehealth covering telecommunications, physical assessment, diagnosis, cameras, and sensors. All telehealth systems rely on a good wireless communication network for data delivery. Other equipment involved will depend on the specific applications, examples include:

Cardiology: stethoscopes, cardiac ultrasound and ECG monitors.

Radiology: probes, MRI and x-ray scanners.

Ophthalmology: retinal cameras, ophthalmoscopes, pachymeters and keratometers.

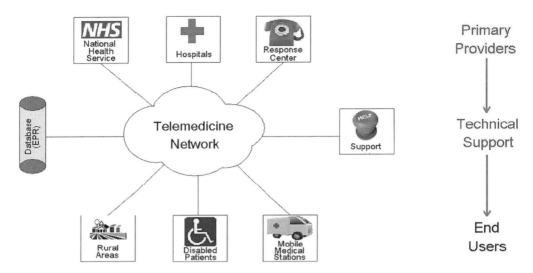

Figure 8.3   Generalized telecare network

Otology, laryngology and rhinology: otoscope, endoscope, laryngoscope and rhynoscope.

Dermatology: dermascope and autoclaves.

To make telehealth more accessible, telehealth services are often delivered through low cost monitoring devices. Another major part of telehealth equipment is a computer server that captures all data through health assessment and is also used to update the electronic patient record system.

A telecare network can be generalized as Figure 8.3, where remote care is provided to end users by various entities via a telemedicine network. Here, the term *telemedicine* network is used with clear distinction from *telecare* since the same network can be used for other medical applications sharing the same communication system. Primary providers would have all the medical equipment listed above, adequate for providing all types of services. A *response centre*, usually a regional hospital with expertise in many areas, is in charge of clinical support and advisory related matters to all *request centres*, including end-users and rural/mobile stations. Technical support is provided by people who look after system maintenance and requires mainly diagnosis and network management and monitoring tools to ensure network availability and data integrity. The telemedicine network, which is a complex communication system by itself as shown in the logical diagram linking various entities together in Figure 8.4, consists of equipment for data acquisition, storage, and transmission of all medical data across a multipoint-to-multipoint network infrastructure. Procedures for database maintenance to ensure electronic patient records (ERPs) are properly in place and data security assurance is also taken care of by technical support people. A wide range of biosensors and remote patient monitoring devices will be installed, either temporarily or on a permanent basis, at end users' sites and mobile support centres. Although there are too many different types of equipment involved for providing support to different applications, in sections 8.3 and 8.4 we shall look

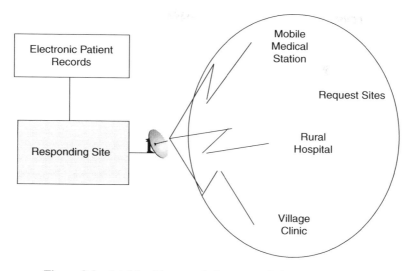

Figure 8.4 A telehealth network that serves different request sites

at one case study each in serving rural areas and elderly care. From there, we can learn more about specific requirements for different telecare applications.

8.1.3   Sensory Therapy

Here, we are looking at senses of the human body, not the biosensors that are attached to healthcare devices. Telecare is also available for those who suffer from sensory and cognitive impairments. It provides stimulations of intellectual activities of five senses, namely vision, hearing, taste, smell, and touch. Multimedia technology allows interactive applications to be built for different kinds of therapies to heal or simply to enjoy a relaxing environment. Multimedia facilitates interactivity through audio/visual (AV) and haptic sensing (see section 9.4 for details) for the eye, ear, and hand. In order to support these multimedia services, a system needs to have a vast amount of bandwidth to provide sufficient QoS assurance (Vergados, 2007). Data traffic requirements, indoor or outdoor propagation characteristics, and network structure need to be taken into consideration when designing a multimedia healthcare system.

This leads to the interesting topic of music therapy that has become popular since the late eighteenth century. It is primarily applied in relieving pain perception to improve a patient's physiological and cognitive state (Standley, 1994). It is also reported that music has an effect on cardiac output, heart and respiratory rate, blood pressure and circulation, and electrical conductance of tissues. (Scott, 2007) even reported a positive impact on cancer treatment. In addition to healing, music therapy is also used in stress relief. While different individuals have different tastes in music, not all music is suitable for use in therapy. Also, the effectiveness of the same therapy applied to two different persons under identical conditions may differ (Darrow, 2001). Finding what might work for an individual is difficult as this may involve some kind of trial-and-error experiment. Irregular physiological response may produce adverse

Electroencephalography (EEG) pattern. Generally speaking, music with a slow rhythm (slower than the nominal heart rate of about 70 beats per minute) tends to be more effective; whereas faster music is often served as an effective stimulus.

8.1.4   Are We Ready?

Telecare is by no means a new area in healthcare services. (eHealth Europe, 2009) has reported that the Spanish Fundacion Andaluza de Servicios Sociales (FASS) of the Junta de Andalucia, being the 100 000th telecare deployment, supports senior citizens of the southern Spain areas with supporting stations situated in Malaga and Seville. It involves a number of entities under the co-ordination of the Ministry for Equality and Social Welfare of the Autonomous Government of Andalucia. More issues exist when more entities are involved.

To promote telecare, there are certain fundamental issues that need to be addressed. First, for sparsely populated areas where no more than a few dozen residents live, there may be no existing network infrastructure available, other basic supporting resources may also be scarce. Another issue, in the context of deployment considerations in continental Europe, is the lack of standardization such that current international standards may not support electronic patient records in certain languages. Some widely accepted international health data standards such as HL7, DICOM, and SNOMED, and American's Health Insurance Portability and Accountability Act (HIPAA), are widely used clinical data standards throughout the world. These cover regulatory requirements, privacy rules, standards and recommendations for implementation. However, these standards are based on the English language and they cannot be applied directly to other languages without some kind of translation. Little incentives exist for practitioners to manually convert patient data into English for the sole purpose of data entry unless mechanisms are in place for direct entry in the language of cocenrn. For example, FASS does have support for information in Spanish. Nationwide implementation is reasonably straightforward provided that the entire system is developed in one single language. However, any attempt to cover countries across Europe may require multi-lingual support.

By asking ourselves the question 'Are we ready for telecare?' The first issue that we are dealing with is who will be the overall in-charge and who will be responsible in the event of something happening.

8.1.5   Liability

Telecare works on the basis that people in different locations are involved in serving end-users. The wider the area telecare covers, the more complex the system will be. Local authorities can supervise everything within a city, when coverage involves areas overseen by different authorities there may be issues such as state versus provincial ownership open for negotiation. To provide comprehensive telecare support, the following entities each with its own organizational structure may be involved:

- Hospitals and Clinics: providing advice and treatment.
- Pharmacies: supplying medication and other medical resources.
- Government Agencies: policies and administrations.
- Medical schools, Public and Corporate Research Centres: research and development.

- Equipment Manufacturers: including but not limited to medical devices and sensors, telecommunications, computers, data storage, etc.
- Telecommunication Service Providers: provide and maintain communication links to connect various entities together for secure and reliable data transfer between them.
- Health Insurance Companies: claims and payouts.
- Patients and End-Users: citizens on both temporary and long-term care and/or monitoring.

Legal issues may also arise in the event of a system failure or when no attention is paid to an event. Failures can occur almost anywhere throughout the telecare system. For example, failure to attend to a situation may be caused by sensor or equipment failure; a network outage or a cable being damaged by workers when maintenance work is carried out. Legal disputes may arise in telecare practices and who should be held responsible in the event of a mishap can be a problematic and complicated issue to address.

8.2   Safeguarding the Elderly and the Aging Population

Aging population is an increasing problem in most developed nations where a higher proportion of citizens are elderly people. The current trend of population aging will certainly lead to shortage of caregivers as well as funding for elderly care in the next one to two decades. As a direct result, deterioration in quality of elderly care can be expected unless something is done in the near future. Although the Nobel Prize in Physiology or Medicine 2009 was awarded to three scientists for their contribution on effects on cell aging (Nobel Prize, 2009), present genetic engineering technologies may still be far from adequate in stopping or even reversing the body aging process. Before this can be achieved, perhaps in the distant future, more people will reach their retirement age in the foreseeable future.

According to the UK Snapshot Neighbourhood Economy Census About ONS Jobs, published by the Office for National Statistics on 7 October 2009, it puts the estimate of senior citizens aged 65 and over by 2033 to 23% of the entire population; compared to only 18% of those aged 16 or younger. Looking at the screenshot of (National Statistics, 2009) in Figure 8.5, the aging population trend in the UK is obviously on an increase. The situation in the Americas is not much better, the US Administration on Aging puts their estimate at 20% of the US population will be aged 65 or over by the year 2030, rising steadily from 12.4% back in 2006. Statistics in Canada also show a very similar picture with an increase of 65+ to 23.4% for the next 25 years from 13.7% in 2006. The chart in Figure 8.6 shows the same alarming yet consistent trend among G8 nations with Japan being projected as the country most affected. The financial burden on supporting national healthcare services will certainly increase (Denton, 2002) over the foreseeable future. Given the severity of population aging and its impact on the society, we must explore how technology can relieve its impacts.

8.2.1   Telecare for Senior Citizens

Telecare, although not intended to provide a preventive solution for senior citizens, can improve efficiency and cost effectiveness to serve the elderly. In this section, we take a look at an example where information and communication technology (ICT) solution is implemented. This enables caregivers to remotely monitor the well-being of elderly people. To help senior

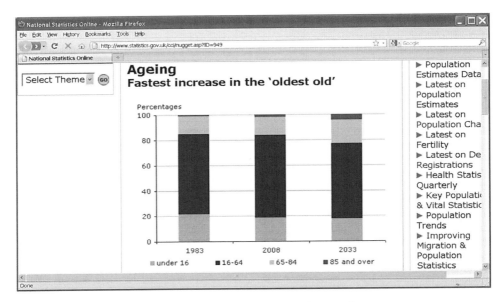

Figure 8.5   Screen shot of statistics on aging

citizens with their daily tasks and make them feel safer while leaving home, a system that serves as an electronic guard by utilizing wireless communications technology can help elderly users stay connected. A generic wearable device that provides advance alerts in anticipation of potential dangers and to remind the user of certain routine tasks is offered. The system can be customized to suit individual needs based on budget and circumstances. For example, a user

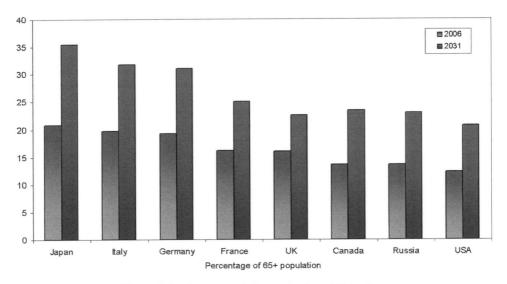

Figure 8.6   Aging population projection of G8 nations

who is at higher risk of falling over can be equipped with accelerometers that automatically detect a fall and send an alert to the caregiver for immediate attention; a user suffering from any form of cognitive impairment can be reminded of various tasks such as flushing the toilet after use, washing one's hands, and taking prescribed medicine at certain times.

Based on an off-the-shelf mobile phone, it provides an inexpensive means of helping the elderly with caring and information gathering by making use of a wearable device and existing wireless communication systems. To the caregivers, ranging from rest home to offsite remote support, they can get easy updates on a user's conditions and receive warnings on emergency situations; making remote monitoring more efficient. To the elderly users, they can be assured that they are well looked after. In case an accident occurs emergency response will be offered. Their health conditions will be monitored. A guard is readily available 24/7 and reminds them about various tasks. They can also obtain advice on demand. Their safety and well-being can therefore be assured.

To provide comprehensive telecare services linking elderly users to their caregivers, the system consists of two separate modules. Each operates independently and linked together through a backbone 3G cellular wireless network. The system block diagram is shown in Figure 8.7, where the caregiver's side is responsible for tasks such as customization of end-user device and acting as a response centre. Whereas the end-user side, namely the elderly user's home, can be as simple as just a pre-programmed mobile phone, to a sophisticated

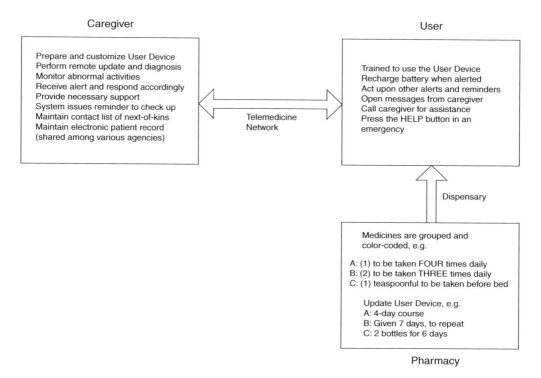

Figure 8.7   Telehealth for elderly care

system with comprehensive features that serves users with special needs such as those with chronic diseases.

System layout is reasonably simple where the user only needs to undergo an informal training that introduces its features and how to respond to various types of alerts and messages. They also need to learn how to seek help under certain situations. More on the benefits to the users will be described in an example below.

The system relies on co-operation from pharmacies when preparing drugs such that a colour coded system can be developed. Medications will be packed with appropriate colour bags. To serve this purpose, one solution that demands minimal effort would be colour printing of labels, so that the drug's name can be printed using a specific colour according to its quantity and frequency of intake. Medication information can also be updated to the user device by a Bluetooth link. The information can be embedded to the prescription so that the device can remind the user of when to take the medications.

For the caregiver, a number of supporting tasks need to be maintained, one of the design motives is to minimize the efforts of the elderly users so that most maintenance related matters are dealt with by the support centre. Prior to delivery, the user device will be fully programmed and configured for its specific functions based on the individual user's needs. A number of separate modules (software or accompanied attachment devices) will be installed during the configuration process.

Once the device is prepared, the remaining tasks will be reasonably similar to the regular duties of caregivers such as nurses and social workers. The system is design to assist them with a wide range of these tasks, including reminder of activities, remote check up in lieu of site visits on certain occasions, automatically alerted to a situation, some necessary support such as consultation can be provided remotely, remotely capturing data for analysis or archival, for example, recovery progress tracking, electrocardiography (ECG) of users either diagnosed with cardiopathy or classified as high risk can be monitored and any abnormal activities identified. To minimize the response time in the event of an accident, the system is designed to remotely detect situations such as a fall that can immediately trigger an alarm, this feature is particularly helpful in nursing homes where elderly people may wander around unsupervised.

This system's modular design can be easily tailored for users with different needs and budget. Implementation can be as simple as a digital assistant that provides a convenient communication link to the caregiver and serve as a reminder for a variety of tasks. Modules can be fitted into the systems for permanent monitoring or certain parts will be installed on a temporary basis to serve certain immediate needs, such as rehabilitation or illness.

The system functional diagram is shown in Figure 8.8, which is capable of serving a user with dementia who is recovering at home after an operation. This particular system consists of both temporarily and permanently installed sensors and equipment that is installed for post-surgical rehabilitation. This example shows the following permanently available features with a central control console enabling smart home technology:

- Thermometer to regular ambient temperature.
- Smoke detector for fire hazard.
- Gas sensor to ensure safe use of stove and reminds user of activation.
- Medication console to ensure prescribed medicines are taken on time.
- First aid kit with reminders for replenishment and alert for expiry.

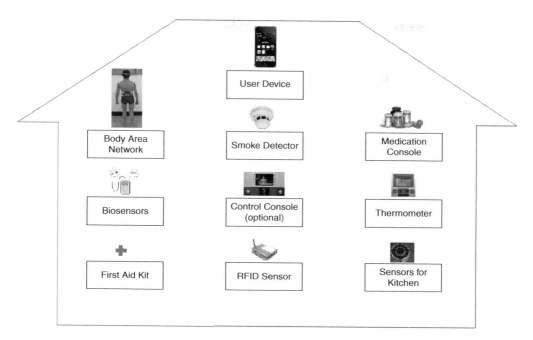

Figure 8.8   Elderly assistive home

In addition to these, one key design feature is the incorporation of smart clothing technology (see section 9.3 for details) where a body area network, as shown in Figure 8.9, is set up with biosensors and accelerometers to detect movement and activities so that various signs related to the user's health state can be collected. Different features can be added based on individual needs. In this example, sensors embedded on clothing can monitor a range of parameters including ECG, temperature, blood sugar level, and pulse rate. Also, should the user fall the response centre will be automatically alerted.

Radio frequency identification (RFID) readers can be installed for a variety of features. For example, when used in the medication console users can be tracked and the medication being taken and when repeat or replenishment should be sought are all recorded so that a reminder will be issued to the user at an appropriate time. A reader that is installed at the door can remind the user to bring the keys and to lock the door. The reader, of course, can also be programmed to automatically lock the door after sensing the user leaving home as an added security feature.

At each user's end, there will be a standalone device that stimulates users with various control methods. Audio command, particularly for users with dementia condition, may require filtering and synthesis to make the speech recognizable. Mobility is also an important consideration as the current system is primarily designed for users remaining at home. The intention of this system is to serve as a companion that can be easily carried with. As the device features a communication module that makes the user feel cared for while away from home,

This system incurs minimal set up costs to the elderly user as the user device can be accomplished by customizing an existing mobile phone. Specific software applications are installed to support a wide range of tasks at the user end. The device will be pre-programmed

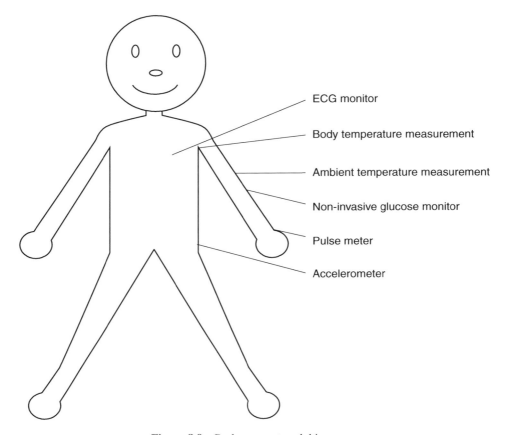

Figure 8.9   Body area network biosensors

for the user depending on individual needs. The Java-enabled portable device is used to serve as a personalized assistant and a monitor. With Java programming, virtually any modern mobile phone can be used for remote assistive care. One suitable mobile phone that can be used by an elderly person requiring care can be bought from the open market for as little as £50 (around US$ 70). While one of the main design objectives is to minimize user interaction so that most tasks are performed automatically, primitive regular operations like battery recharging need to be taken care of by the elderly user. Since most mobile phones currently available make use of micro-SD memory cards, it is possible to roll out feature enhancements and to provide software update by swapping over the installed card for re-programming. Other functions such as continual diagnosis of both the user's well-being and the system's condition can be performed remotely.

To illustrate how flexible the device can be, we refer to the device in Figure 8.10 where the following functions are supported by a 3G mobile phone with touch screen user interface. Although a touch screen makes it easy for use by the elderly, this is not a basic requirement to enjoy basic telecare services brought to them by wireless technologies. This particular device has the following features:

Figure 8.10   Assistive care mobile phone

- All basic functions of a mobile phone, with message filtering capability that sorts only messages from the caregivers into its default mailbox, a help button for real-time support is also provided.
- Checking the user's health, analysis of captured vital signs such as heart and respiratory rates, blood glucose levels, etc.
- Suggest what to eat for the next meal, based on nutrition balance and any existing medical conditions, can be linked to an online ordering system for home delivery, similar to that of some paid-TV subscription set top boxes.
- Link to a home medication console where medicine is stored, the colour coded system described above is implemented to assist with ensuring appropriate time and quantity of each medicine taken. An RFID system would keep track of medicine taken and remaining stock.
- The RFID reader featured in Figure 8.8 is intended to serve as a door guard at the main entrance of the user's home. Its main purpose is to ensure that the user has not forgotten the keys before leaving home, once the user unlocks the door when the keys are not removed an alarm will be generated. Similarly, the system will also remind the user to lock the door securely before leaving. An automatic locking mechanism can also be engaged.
- Entertainment features are also available. In this illustration, these two options are subdued since they are auxiliary functions that are less important than healthcare related support functions. Music also serves as an effective tool to calm the user down when agitated;

memory games that stimulate the user can keep the user engaged in some brain-training activities.

The flexibility of Java allows many other services to be included with appropriate networking sensors. These can include fall detection, stroke detection, body temperature monitoring, prognostics of implanted medical devices, etc. Built upon open-source platform that is not restricted by licensing issues, applications serving different needs can be developed and installed on a wide range of generic off-the-shelf portable devices.

User-friendliness is an important design consideration since most senior citizens are not familiar with technology. Another major function is to collect information about users' health conditions such as blood pressure, body temperature and SpO_2 readings, medication and nutritional intake, and fall history. Such clinical information will be analyzed on a regular basis for monitoring purposes. In addition, the clinical information can be connected to and shared with healthcare facilities (e.g., general practitioners or hospitals) using an existing wireless network. This feature is particularly suitable for older adults with cognitive impairment and users who are recovering at home after hospitalization (after hip fracture surgery) while still under close surveillance by hospital staff. In addition, this feature can help reduce demands in hospital resources as well as travel time for elderly patients.

8.2.2   The User Interface

Routine activities of senior citizens can be supported by a multi-sensory telecare system as an electronic guard. Elderly users with special needs such as memory loss and cognitive impairment sufferers can be greatly benefited by technological advancements in human-computer interaction (HCI) and wireless communications. A wearable therapeutic device provides general assistance, health monitoring, calling for emergency assistance, alerts and reminders; can provide dementia sufferer with a peace of mind. This solution also links care providers and elderly people together, particularly those living alone, so that they can stay in touch.

The HCI interface, that determines how user-friendly a device is, must be very carefully designed. In particular, attention must be paid to ensure elderly users will find it easy to operate. HCI involves consideration of the following:

- Language.
- Engineering feasibility and cost effectiveness.
- Mechanical reliability and durability.
- Precision.
- Ergonomics and human factors.
- Cognitive psychology and sociology.
- Ethnography.

There are virtually infinite choices of implementation methods including keypad, computer mouse, touch screen, navigation menus, etc. In dealing with user interface design, we must mention Shneiderman's 'Eight Golden Rules of Dialog Design' (Shneiderman, 2005). This set of rules describes:

1. Strive for consistency
   - Consistent sequences of actions should be required in similar situations.
   - Consistent styling with colour, layout, capitalization and fonts should be used throughout.
   - Identical terminology should be used in prompts, menus, and help screens.
2. Enable frequent users to use shortcuts
   - Abbreviations, special keys, hidden commands, and macros can be assigned to increase the efficiency of interaction.
3. Offer informative feedback
   - The system should respond in some way for every user action so that the user knows the input has been collected.
4. Design dialogs to yield closure
   - Sequences of actions should be organized into groups with a beginning, middle, and end. Provide instructive feedback at the completion of a group of actions confirms command execution.
5. Offer error prevention and simple error handling
   - Forms to be organized in such a way that obvious errors will be disallowed, caution should be exercised to accept certain exceptions, e.g. Telephone entry may include characters such as '+', '−', and brackets for area codes.
   - Instructions should be offered upon detection of an error and to offer simple, constructive, and specific instructions for correction.
   - Segment long forms and process each section separately such that any error will not cause total loss of information already entered.
6. Permit easy reversal of actions
   - Let users go back through menus.
7. Support internal locus of control
   - User override and manual intervention. Must assure ease of information retrieval and avoid monotonous data entry sequences.
8. Reduce short-term memory load
   - Theory suggests that a typical person can store something between five and nine pieces of information for short term. One can relief short term memory load by designing screens with clearly perceptible options or using pull-down menus and icons that lists out every available option to avoid the need of memorizing.

As a final note, operational reliability depends heavily on prevention of errors whenever possible. Necessary actions can be taken in user interface design in such a way that error occurrence is minimized by using methods such as organizing screens and menus functionally; and designing screens to be distinctive thereby making it almost impossible for users to mistakenly carry out irreversible actions that may cause data loss or system malfunction. Understand target users, particularly when designing systems for elderly users, should expect users to make mistakes or inappropriate entries, special attention should be paid to look ahead to where users may make mistakes, so user interface design can preemptively take these into account. Exception handling would prevent unpredictable system response in the event of user executing an invalid command.

8.2.3   Active Versus Responsive

As a reminder to the readers, telecare is not intended to prevent accidents from happening. For example, telecare systems do not have the ability to counter-balance the user in the event of falling over. Such systems operate more reactively as certain rules are programmed to respond to different scenarios. There are telecare devices that are more responsive to assist with preventive care. Any such device helps stimulate users so that they are trained to keep themselves engaged in some kind of activities.

Although 'prevention is better than cure' may be heard countless times, accidents do happen despite all best practices being in place. While technology can sometimes prevent an accident from happening, technical solutions are far more often passive than active. As in the case of modern motor vehicles where many safety features are built in to enhance safety, many of them only reduce the risk of an accident from happening or minimize the impact of an accident; many of these technical features do not have the ability to stop an accident from happening. For example, parking distance control (PDC) automatically alerts the driver when coming close to a physical obstacle. However, it does not apply brakes to stop the vehicle from bumping into an obstacle. Collision can only be avoided if the driver stops the vehicle manually. Similarly, telecare technology is mainly responsive as many of those featured in Figure 8.8. Very few systems have the capability of actively pre-venting an accident by proactively performing a task upon early detection of hazardous activities.

Most active telecare systems involve artificial intelligence. For example, detection and analysis of daily activities so that early signs of warning can be generated before something serious happens. This is particularly useful for elderly users since what they see and what their inner ears sense may sometimes differ. In theory, any proactive system should address such differences and initiate corrective actions before something goes wrong. For example, a fall can be prevented if an imbalance is detected so that counter balance can be activated prior to an actual fall. An elderly person may see the ground and senses how to move across by avoiding surrounding hazards, but the actual action taken when taking a step forward may differ from that perceived. This happens because visual reference may be distorted due to poor eyesight. Counter balancing such difference can be accomplished in a similar way as a cruise ship stabilizer that reduces the effect of rocking motion. Stabilizers keep the ship straight and upright in waves and adverse weather conditions (Dear, 2006). These stabilizers function by extending wing-like flaps on both sides of the ship. As the ship sways the stabilizing mechanism will counteract by exerting a force, through distribution of its own weight, in the opposite direction so that it will maintain a good balance. In the context of telecare, an equivalent system takes into account the user's life habits by constantly monitoring what the user does. The system can be 'trained' to respond to any abnormalities to initiate responsive actions.

In addition to these existing solutions that provide assistive elderly care, there are also other implementation options such as using a set top box based solution for providing monitoring and information on healthcare via the TV remote control, linking to information services but also directly to telemedicine and security equipment such as monitors and wireless cameras (Scott, 2009). The system is scheduled to commence trial around the first publication of this text in 2010. In section 9.2, we shall look at the system in more detail as an emerging technology solution.

8.3   Telemedicine in Physiotherapy

Physiotherapy, or physical therapy, is widely practised to relieve deterioration in movement due to aging (Geriatric), injury (Cardiopulmonary and Orthopedic), or disease (Neurological). As physical movement involves areas such as limbs and the back, which commonly relates to biomechanics of joints and spinal manipulation, sensors may need to be small and able to detect minute movements in 3-D space.

8.3.1   *Movement Detection*

As recovery and progress monitoring involves detection of movements, there are two major methods, namely sensors and video analysis. Many sensory systems are infrared based where movement of infrared emitting sources, such as human bodies, are tracked; others involve mechanical switches and sensors, such as accelerometers and vibration sensing. Movement of different parts of the body may require different mechanisms. For example, spinal curvature (Chow, 2007) has very different requirement when compared to knee position (Brinker, 1999). Regardless of which technology is used, there is always a trade-off between coverage area and precision.

Video sensing, as illustrated in Figure 8.11, can easily track the movement of the entire body inside a confined area. The coverage area depends on camera placement and lens' focal length. In this example, six cameras, as shown in Figure 8.12, are installed. All cameras are connected to the computer, either with cable or wireless, so that the image captured from each camera at a given point in time can be compared and analyzed. By comparing the images acquired by all cameras with those of adjacent frames, movements can be tracked. The camera in Figure 8.12 has a photographic lens mounted, just as those used in single-lens reflex (SLR) cameras. The longer the lens' focal length, the more detail is captured with a more close-up view so that greater precision is yielded. However, the angle of coverage is also reduced. Conversely, a wide-angle lens provides wider coverage at the expense of less detail and precision. Such

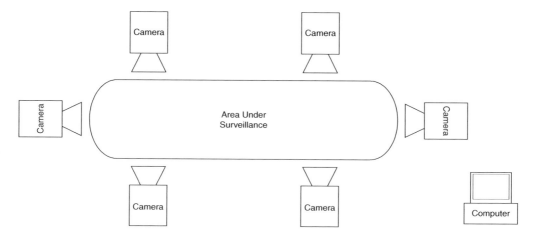

Figure 8.11   Video motion sensing network

Figure 8.12   Motion tracking camera

trade-off is illustrated in Figure 8.13 where the image on the right is taken with a focal length three-times that of the left.

A widely used alternative is accelerometer network. By placing a number of accelerometers on the subject's body, such as the dummy shown in Figure 8.14; when the subject moves, each accelerometer will sense the movement in each of the three dimensions. Although the example in Figure 8.14 connects each accelerometer together using wires, additional circuitry can be installed with Zigbee communication capabilities. For details on the standards governing communications between medical devices, please refer to the Appendix. An accelerometer is low cost and simple to fit, the one depicted in Figure 8.15 is capable of detecting 3-D movement. When fitted, any movement can be sensed and sudden acceleration (change in speed and or direction) that may indicate a fall can trigger a remote alarm.

All these rely on technologies for detecting the magnitude and orientation, direction and speed of movement. One major drawback of using accelerator for fall detection is that it operates by measuring its acceleration relative to freefall due to gravity. Therefore, an accelerometer will not produce an output when it undergoes freefall. To combat this problem, an accelerometer should be installed at an offset angle that produces a relative movement with respect to its vertical axis while falling downwards.

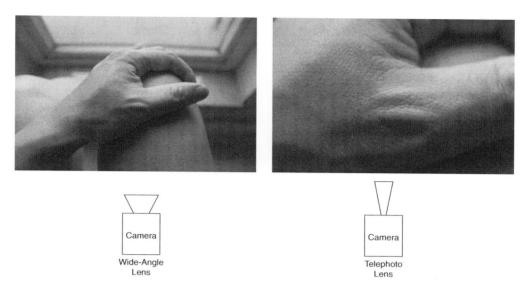

Figure 8.13   Lens focal length versus coverage angle

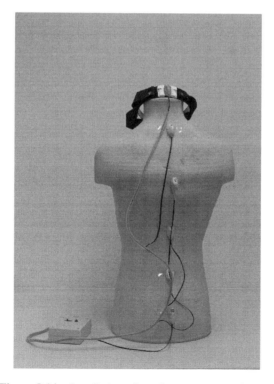

Figure 8.14   Installation of accelerometers on a dummy

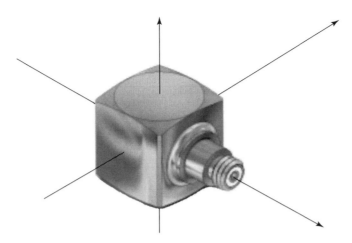

Figure 8.15   An accelerometer senses 3-D movement

8.3.2   *Physical Medicine and Rehabilitation*

Physical medicine and rehabilitation, also known as physiatry, is aimed at regaining functional abilities in an effort to combat the impacts on disabilities. It deals with the recovery of muscles, bones, tissues, and nervous systems. Prognosis for various neuromuscular disorders can be accomplished by nerve conduction studies (NCS) and needle electromyography (EMG). As NCS involves electrical stimulation to peripheral nerves, these can be conducted remotely so that the patient does not have to travel to the clinic for diagnosis. This is particularly suited for spinal cord monitoring and used in studies on the impacts of schoolbag on children's back (Chow, 2006). We shall look at a case study involving prevention of spinal injury by studying the effects of weight distribution of schoolbags on children in a case study in sub-section 8.3.3 below. Before this, we continue to look at how technology advances in telemedicine can bring relief to patients with physical impairments.

Palliative care and rehabilitation have long been considered as two important parts of comprehensive medical care for patients with advanced disease (Santiago-Palma, 2001). It also suggests that physical function and independence are, as in the case of elderly, important attributes for both patients and than caregivers. Palliative care involves psychological and spiritual support as a means of relieving distressing symptoms. Regulations governing the application of palliative care may differ in different countries. For example, the US requires certification by two physicians for a terminally ill patient whose remaining life expectancy is less than six months to be eligible for enrolment (Sheehan, 2003). No such regulation exists in most other countries.

8.3.3   *Active Prevention*

Although telecare does not normally deal with prevention, technology does provide mechanism for active prevention. For example, a patient who exercises after knee arthroscopy may need to restrict the amount of movement to prevent causing further injury in the event of overstretching. Necessary actions such as controlled passive stretching, hold–relax, repeated contractions and assisted active exercises may be necessary for the recovering limb and free active exercise for unaffected areas so to reduce edema. This does not consolidate and cause joint stiffness. Patellar tracking would become necessary for ensuring speedy recovery (Brunet, 2003). The appropriate installation of accelerometers would detect early signs of movements that may cause contractures and deformities, essentially serving as a splint that has the capability of dynamically tracking movement, instead of using a traditional static splint that immobilizes the entire limb. Limited range of movement is therefore allowed without the risk of overstretching. There is, however, a small catch with sensor placement since they must be installed without nerve compression. Also, the force exerted to the sensors may be reduced by padding for bony prominences or areas where the bones protrude slightly below the skin. There are two important points to remember when placing sensors. First, the sensors themselves should only detect movement specific to the limb, but not the vibration that may be caused when the patient walks. To compensate for vibration, prognostics techniques may be necessary for the associated electronic components (Gu, 2009). Prognostics will be discussed in section 9.1.

Another consideration is wireless transmission of captured data. How to respond to a sudden situation and how to ensure no critical event is missed. A mechanism for ensuring continual communication link availability may be necessary for each sensor and its networking device. For example, a polling system that sequentially checks the readiness of each sensor would ensure that all sensors are in range. A controller must be pre-programmed to detect early signs of a possible risk. This may involve the implementation of *fuzzy logic*, an 'intelligent' problem solving algorithm installed in an embedded system. The key feature of fuzzy logic is the ability to derive a decision based on equivocal and incomplete information. In this context, the algorithm is capable of detecting an alarming situation prior to its occurrence, based on subtle abnormal signs.

Unlike simple embedded system controllers that execute basic response based on a number of pre-defined parameters such as in the simple case of wireless insulin control system in Figure 8.16, where a simple controller regulates the glucose level by feedback from the

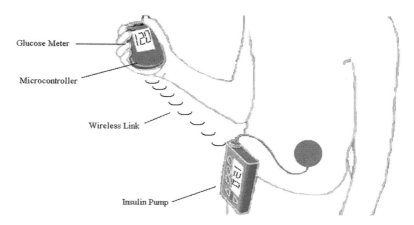

Figure 8.16   Wireless insulin pump

glucose meter that controls the amount of insulin pumped solely upon the meter's reading. In a fuzzy logic controller, the algorithm relies on picking up the rate of change of reading captured from the sensors connected to it such that it responds to the detected changes. For example, an array of accelerometers installed to detect the fall of an elderly patient may rely on successive readings that exhibit a significant change in movement relative to the regular pattern being read while walking steadily. The sudden significant change in reading within a relatively short period of time may be many times greater than that of what normal activities will generate. These readings do not have to follow any logical pattern in order to be identified as the detection of a fall so that it is not necessary to tweak the reading into any logical description.

Fuzzy logic implementation involves defining the control criteria and parameters. In the example of fall detection the parameters would be readings obtained from individual accelerometers. What is the normal range of reading when the patient undertakes normal activities? Are all sensors experiencing the same readings? What are the input and output relationships, does simultaneous detection of sudden acceleration downwards indicate a fall, or the patient intentionally bend down? The rule based nature entails a series of expression:

$$IF\ X\ AND\ Y\ THEN\ Z$$

that collectively define the output response for the given set of input conditions, namely the X's and Y's of each expression in the series. The simultaneous occurrence of X and Y would trigger the corresponding predefined action Z.

Remember, one objective of implementing fuzzy logic is not only to detect the occurrence of an event but also to proactively warn in advance the risk of an event. So, an output should be generated to indicate the risk when something is detected prior to its happening. An imbalance that may lead to a fall should therefore be detected and warning issued prior to an actual fall. This will be triggered by a set of abnormal readings, such as the situation where X suddenly rises while Y descents. Any possible preventive actions can therefore be activated. Other expression may include consequences of post event action such as automatically alerting a response centre after a fall.

8.4   Healthcare Access for Rural Areas

The problems associated with providing healthcare services in rural areas are very different than those in urban areas. Rural residents face a unique combination of factors that create disparities in healthcare. Lack of recognition by legislators and the isolation of living in remote rural areas are perceived by policy makers as money not worth spending. In isolated areas, where residents are more likely either self-employed or retired, are far less likely to enjoy employer-provided healthcare coverage.

Funding is one major issue in any national healthcare system (Roemer, 1993), any vast project in extending healthcare services must therefore produce observable return-on-investment (ROI). Providing healthcare services to rural areas can be a significant challenge because of the population density that makes support very expensive. To demonstrate the connotation of serving rural areas, we take a look at the case study in the US. According to the American Hospital Association (AHA), rural hospitals serve 54 million rural American residents. This magnitude is comparable to the entire UK population. First, the financial incentive is an issue for operators. Amongst the 54 million people living in rural areas, there are nine million Medicare beneficiaries. Medicare margins are lower particularly with small hospitals. Justification for the establishment of any adequately equipped hospitals would be difficult from a financial point of view. Driving down the cost of providing healthcare services would increase profit margin for service providers, this can be accomplished by advances in healthcare technologies and simplifying processes and formalities, this can extend medical services to rural areas more efficiently and cheaply.

Another major problem arises from accident recovery that leads to prolonged delay between an accident and response. Many of these delays are related to increased travel distances in rural areas and personnel distribution across response centres. In response to these problems, the US government's Telemedicine Report to Congress (Kantor, 1997) states: 'Telemedicine also has the potential to improve the delivery of health care in America by bringing a wider range of services such as radiology, mental health services, and dermatology to underserved communities and individuals in both urban and rural areas', acknowledging the importance of providing healthcare services to rural areas through telemedicine.

Telecare is particularly suitable for rural areas where people can live alone with the assurance that they are well looked after. It has the following key features:

- Bring medical and healthcare technology to people everywhere.
- Provide more efficient and affordable services with easy access.
- Healthcare services are no longer limited to certain locations such as clinics and hospitals.
- To assist disabled people, take care of the children and the elderly, heal the sick or injured, and to support vulnerable individuals.

There are, however, certain pre-requisites that need to be dealt with. First, supporting infrastructure that provides coverage to the areas of concern must be available. For example, an existing wireless network with sufficient bandwidth that can support all necessary healthcare services. As telecare involves people in various locations, liability issues must be thoroughly addressed before providing any remote services. In this context, we may need to ask questions like who is responsible for overseeing the process, what if a mishap leads to liability related issues, what happens if an accident causes injury or fatality and who would be held liable, etc. All these decisions and liabilities related questions need to be properly documented.

One main deployment consideration is whether existing infrastructure, if any, can support the desired services in terms of providing adequate resources and geographical coverage. In vast areas with low population densities, there may be no support at all; small settlements may have only very primitive telecommunication networks such as the plain old telephone system (POTS) available for nothing more than voice calls. Serving the farming community may be even more challenging because the houses can be several miles apart. Even a small local clinic with the most basic equipment can be difficult since there are perhaps only a dozen of people within its proximity. Providing wireless telecare services is extremely difficult because of excessive signal loss. Going back to the fundamental issue of existing infrastructure again, the lack of adequate networking resources is even more acute in developing countries. Cloud computing is becoming increasingly popular in recent years. It may change the way IT infrastructure advances, this is likely going to be particularly helpful for developing countries (Cleverley, 2009). According to the definition on *wiki*, the 'cloud' is a metaphor for the

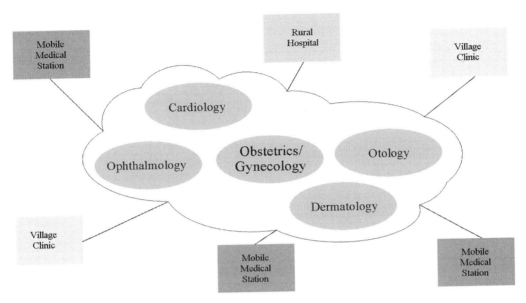

Figure 8.17 Telecare network

Internet; such that it can support virtually any kind of services, including a range of healthcare services. Cloud computing, originally developed as a platform for supporting various common business applications online that are accessed from a web browser, aims at making over the ICT model from a largely static connection between applications and hardware, and discrete expansion dictated by physical equipment limitations to an integrated computing platform capable of more granular scalability and flexibility. Such deployment freedom supports a wide range of multimedia telecare services operated by different entities through different modes of application delivery. Adopting the basic cloud computing conceptual model, it can provide a number of telecare services as shown in Figure 8.17.

Let us take a look at a case study in Grainger County, Tennessee of the US, a rural area with approximately 20 000 residents without a hospital. This area is geographically isolated with limited road access due to obstruction by the Clinch Mountain and the surrounding lakes. The project 'Rural Health Care Through Telemedicine: An Interdisciplinary Approach' has been implemented by the US Office of Rural Health Policy Rural Telemedicine Grant Program and the University of Tennessee with the main objective of improving access to healthcare services and to reduce the isolation of service providers in the county. Each of the county's four clinics had an interactive audio-video telemedicine system installed and clinicians were trained to use the system for patient consultations. This supported a primary care physician in one of the rural clinics to examine a patient with remote support by specialist physicians at the university medical centre some distance away. For emergency health services, two EKG units each capable of transmitting 12-lead EKG data from a patient to both the local clinic and remote university medical centre were connected via a mobile phone network. For other non-emergency consultations, patients could communicate with service providers using a video phone connected to the plain old telephone service (POTS) network. This system provided a mechanism for basic healthcare services for residents in a geographically isolated area. This is only possible given the necessary funding that supports initial deployment. To implement similar systems in a rural area, financial feasibility is most likely a major constraint. The readiness of adequate existing network infrastructure and interoperability standards for necessary supporting software may also be issues that need to be addressed.

8.5   Healthcare Technology and the Environment

The industrial revolution changed the landscape of manufacturing and mining in America around the dawn of the nineteenth century, fossil fuel burning and toxic gas discharge have significantly accelerated, that in turn create health related issues such as air pollution and acid rain. Although there is no doubt that industrialization directly causes negative environmental impacts on people's health, the trend of industrialization spreads eastwards into Asia from the early time of the post-war era. For example, the highly insanitary business of battery manufacturing saw its shift from the US to Japan around the 1970s then into China about two decades later. Health hazards associated with industrialization therefore shifts gradually from developed countries to third-world countries where the general sentiment is willing to trade health deprivation for monetary profits. There are close relationships between healthcare and the environment, as well as the technologies behind. We intend to conclude this chapter by taking a look at why healthcare is so closely linked to the environment, how healthcare technology plays a role in environmental protection, and the kind of environments healthcare technology is bringing to us. Healthcare technology has many implications on the environment, everything from pollution of biological waste to radiation that may be hazardous. Conversely, environmental impacts can affect healthcare and technologies related to it. For example, regulatory constraints may prohibit the use of certain materials; environmental impact on disease spread also causes great concern over centuries.

8.5.1   A Long History

The links between healthcare and the environment have been deliberated for centuries. The first reported plague pandemic case was probably that of year 541 originated from Egypt. Better known as 'The Plague of Justinian', it affected much of the Eastern Roman Empire (Little, 2008). It is widely believed that bubonic plague first made its way to Europe through grain ships that had housed an immense rodent population. Believed to have wiped out around half of Europe's population by the year 590 (Maugh, 2002), the plague continued to roam the world for another century before it subdued. Next was the *black death* that haunted much of the world around the middle fourteenth century. It was probably the best known example where healthcare technology and the environment had a very close tie. Some 600 years ago, there were thought to be three types of plagues responsible for wiping out an estimated half of Europe's people. (Kelly, 2005) suggested that the culprit was most likely a viral hemorrhagic fever that was spread out of control by rodents. It was suggested that fleas that carried the plague originated from Asia and rats carried them into Europe by merchant vessels (Cartwright, 2004). When symptoms started to appear, a victim typically had a remaining life expectancy of about a week. Without any defence or knowledge of the cause of the pestilence, physicians were unable to provide any cure so those who got infected were abandoned. The disease

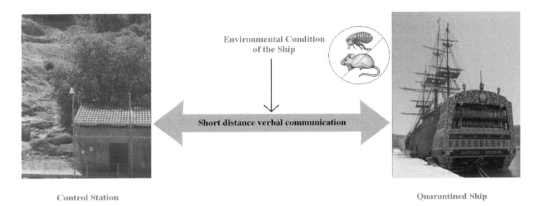

Figure 8.18   An ancient telemedicine system

proliferated very vigorously as victims communicated it to the healthy individuals who came near them.

Any effort in containing the disease must involve knowledge about its spread. Any mechanism that can combat the disease requires information to be gathered. Any such 'technology' was not available during the outbreak. In fact, the causes of plague were not discovered until the nineteenth century. The plague, originally attached to rodents, was believed to be transmitted to humans by fleas. A flea, carrying ingested plague-infected blood from its host, i.e. the rodent, can live for as much as a month away from that host before finding its way to a new host, i.e. a human being. The plague therefore spreads as the flea sucks blood from the human body when it injects into that victim some of the blood already within it. Early telemedicine found its presence when people realized that the spread of the plague could have been contained after a certain period of isolation. Ships suspected to have carried the plague were therefore quarantined and identified with a flag. They were only allowed to dock when the plague was thought to have vanished after the quarantine period. This communication path between the ship and the shore, uses primitive health information communication technology as shown in Figure 8.18, would have been a good example of early telemedicine deployment where information about the environment and the situation inside the ship is sent out for remote investigation. The ship will communicate with the control centre so that rodents and fleas that carry the plague cannot board the shore before the ship's environment can be assured to be safe.

Technology, if available at the time, would have helped in many ways. First, a telemedicine system would have helped diagnose and quarantine those who were infected, providing information on treatment and therefore a better chance of survival. Clusters of those infected could be linked together for information sharing. Also, the plagues' spread pattern could have been analyzed thereby reducing the risk of spreading further by containing it. The study of plague could provide some insights into other types of pandemics. Although we know by now that plagues can be controlled by antibiotics such as streptomycin, gentamicin or tetracycline (Massachusetts Institute of Technology, 2008), some 2 000 deaths from plague are still reported around the world each year. Telemedicine would provide a measure for combating plagues with mobile medical monitoring devices that continuously assessing the health of people in different areas. Such monitoring systems will improve the ability of health authorities in different areas to react to and predict disease outbreak and its epidemiological spread due to different bacteria or virus.

To assess the disease spread pattern and analyze the environmental impact on disease outbreak, computational modelling is regarded as the most appropriate method (Bloch, 2009). A model can be generated by collection of information based on spatial and temporal information of the occurrence of an infection, for example, a progression model that analyzes epidemic spread in homogeneous and heterogeneous networks (Zhang, 2009). Such a model is developed for computing the spread of effects such as environments, climate and people traffic between countries and regions in various scenarios. Information about each reported case is collected for computing the dynamic history of an outbreak to connecting clusters by adding the appropriate demographic and population mobility information. This would expand to a spatially and chronologically structured stochastic disease model that simulates the spread of epidemics from a suspected origin across the globe. For example, the 2009 A(H1N1) swine flu is believed to be originated from Mexico (Centers for Disease Control and Prevention, 2009). To predict its spread pattern, a computational model would commence by using the first set of data for simulation with a cluster of infections that occurred in March 2009. As the disease spreads, the predicted infected areas will expand outwards from its origin. Initially, data collected at the level of individual countries. Spots throughout the world appear due to rapid people movement resulting from air travel. Further cases of outbreak about the location and time would then be appended to construct a more comprehensive model over time.

There are, however, uncertainties involved in modelling how the disease spreads. In the modern world, seemingly random movement of airplanes that bring infected people across the world may carry the disease in a haphazard pattern such that spread phenomena with physics-based knowledge can no longer be used. As with historical events that date back to the Plague of Justinian, the environment plays a vital role in manipulating disease spread. The relationship between environment and disease could also be seen in the spread of diseases as changes to the ecosystems caused by environmental pollutions flourish pathogen growth (Briggs, 2003). Water contamination, poor sanitation and poor hygiene are all contributing factors to rapid disease spread. Environmental protection therefore remains an important factor in disease control.

8.5.2   Energy Conservation and Safety

Energy conservation is always perceived as closely linked to environmental protection because of the general belief that consumption of non-renewable sources impacts the environment. Designing a medical device that is energy efficient is one important step towards maximizing return-on-investment (ROI) and product reliability. Particularly important for mobile medical devices, energy efficiency improves the cost effectiveness of a device and prolonged battery life. Safety related to the use of medical devices is vitally important since a device failure may lead to fatality. Safety assurance may entail:

- Identify potential hazards during operation.
- Quantify damage potential, for example, through computational modeling or prognostics techniques.

- Evaluate all necessary safety measures.
- Take remedial measures for reducing and controlling risks.
- User training to ensure proper use.

Protective housing is always a vital measure to ensure operational safety. However, there may be trade off between efficiency and level of protection related to the material used. For example, although metal housing may provide virtuous protection against physical impact and EMI, metal is generally not a suitable material for medical devices used in telemedicine applications since many of them transmit and receive data through wireless media. With wireless medical diagnosis and monitoring devices, metal is generally not suitable because of its conductive properties that reflect electromagnetic energy at the surface except at extremely high frequencies. Electromagnetic energy penetrates a distance into the metal that increases with wavelength, known as the 'skin depth'. Lower frequency electromagnetic waves can therefore propagate through metal with a certain amount of attenuation if the housing is thin enough, while higher frequency electromagnetic waves, because they do not significantly penetrate into the metal, are reflected just like a mirror. Such properties can also be useful since the conductive housing of the device then effectively shields the internal electronic circuitry from higher frequency electromagnetic interference that could adversely affect device operation.

Careful design consideration is necessary for energy conservation and transmission efficiency if the transmitting antenna is also within the conductive housing. Telemetry, technology that allows remote measurement and reporting of information, must be performed with lower carrier frequencies since the housing effectively acts as a low-pass filter. This reduces the effective data rate that can be supported by the system and increases the necessary transmitting power for an implantable device when sending captured data to a remote device. Such additional transmitting power requirements for the implantable device results in limiting the range over which it can operate.

In these transmitting devices, particularly for implantable devices, efficiency and orientation may impact power consumption. An antenna for such application should be coupled with a reflective plate to increase the gain of an antenna in a selected direction, this normally points away from the patient's body. Electro-magnetic wave radiated by the antenna tends to attenuate when it encounters obstructions such as tissue and water. An antenna that is designed with a selected transmission direction is known as a 'directional antenna'. Its main advantage is to enhance the power of the antenna in the selected direction thereby increasing the transmission distance. Such a type of antenna usually has a reflective plate on one side to increase the directionality and gain of the antenna. The reflective plate reflects radiated signals for transmission as well as reception hence increases the directional radiation gain of the antenna.

In addition to antenna efficiency, control of Specific Absorption Rate (SAR) also needs to be optimized for power saving. Transmitting devices are required to meet certain regulatory requirements for maximum SAR levels in some countries. Such regulations are aimed at imposing appropriate limits for users of wireless devices from the perspective of energy absorption into body tissues. SAR is a description of the time t derivative of the incremental energy dU deposited in an incremental mass dM contained in a volume element dV of density ρ, written as:

$$SAR = \frac{d[dU/dM]}{dt} = \frac{E^2}{\rho} \qquad (8.1)$$

SAR, measured in watts per kilogram (or equivalently milliwatts per gram mW/g), is a measure that estimates the amount of radio frequency power absorbed in a unit mass of body tissue. The SAR restriction varies from around 1.6–2.0 mW/g depending on legislation. Also, SAR limits can be different for different regions of the human body. Compliance with the applicable maximum SAR limits is usually obtained under specific environmental and operational conditions.

Practical SAR values can deviate from anticipated measurement results due to a number of reasons:

- Frequency or energy of the incident radiation relative to the composition of the tissue mass being measured.
- Radiation intensity of the device and the proximity of the device to the tissue.
- Any presence of nearby reflecting surfaces and their orientation.
- Transmission power of the device to establish and maintain communication.
- Orientation (Polarity) of the field vectors relative to the tissue.

All these can be managed during the design stage. However, in order to comply with certain standards the carrier frequency may be fixed. Active control of the transmission power output can optimize battery life. Improving transmission efficiency is also necessary to prevent antenna detuning that can occur from nearby objects due to electromagnetic capacitive or inductive coupling. Shielding by use of appropriate housing or active control of the antenna can combat these problems.

8.5.3 Medical Radiation: Risks, Myths, and Misperceptions

Ever since the discovery of X-ray for medical imaging that relies on the different rates of energy absorption by different tissue (and bone) types, radiation exposures from diagnostic medical examinations have been considered for over a century (Filler, 2009). As discussed in sub-section 4.3.1 earlier, the effectiveness of X-ray radiography is governed by the intensity of radiation dosage. The amount of radiation that may cause health problems need to be thoroughly investigated. (Willforth, 1985) suggested that in America, human exposure to ionizing radiation is almost all related to medical diagnostic radiology, which suggests that radiation comes from the ambient environment. This leads to a question that does have grounds for dispute because *gamma rays* from disintegrating nuclei of radioactive substances that naturally exist, Gamma radiations discharge even more energy than X-ray (Als-Nielsen, 2001).

Therapeutic uses of radiation naturally involve higher exposures. Its associated risk is assessed by a physician before examination. Standardized radiation dose estimates can be given for a number of typical diagnostic medical procedures yet the dosage can be very different depending on individual circumstances. Each patient's metabolism and the type of examination are important consideration factors for the dosage. The exposures are widely considered as comparable to those that are routinely generated from natural radiation in the surrounding environment. Obviously, some energy of the X-ray dose is absorbed within the body since bones and tissues block some of the radiation that in turn forms the radiograph showing up as shadows on the film. As a consequence, some cells may die prematurely although the amount of cell damage is quite minimal. Such damaged cells do not actually pose any risk since they are naturally replaced. However, there is a health risk caused by some of the cells not dying, but instead sustaining genetic damage. Such damage can, in rare cases, result in the cell becoming cancerous.

The dosage depends on applications and diagnosis areas. For example, the typical dosage of dental X-ray is around one-third of that of a chest X-ray. Computed Axial Tomography (CAT) scan, also known as CT scan, subjects the patient to the X-ray scanner for less than 30 minutes to complete a full body scan. Some CAT scanners use up to 300 X-ray scanners taking 300 pictures each. This generates some 90 000 X-ray slices or tomograms to form the overall picture. The amount of radiation received in a CAT scan is usually around 10 mSv, this is equivalent to about 60 medical X-ray doses. Note, incidentally, that this is approximately twice the recommended maximum radiation dose for a pregnant woman. CAT should therefore be avoided for pregnant patients.

As described in section 4.2, positron emission tomography (PET) is a nuclear medicine imaging technique that relies on the circulation of an injected radioactive substance that emits *positrons* (high speed electrons) and *gamma rays* (highly energetic ionized radiation produced by sub-atomic particle interactions). PET relies on detecting pairs of gamma rays emitted indirectly by a radioactive tracer and the scanner reads gamma rays like the CAT scanner reads X-rays. As the radioactive tracer travels through certain parts of the body, PET is capable of producing more detailed images of a specific organ. However, the emission of gamma rays may pose health hazards. With a typical dose somewhat higher than that of a CAT scan, the use of PET should be precluded unless it is absolutely necessary.

Sources of radioactivity in the ambient environment includes the colourless, odourless radioactive noble gas Radon (Rn^{222}); itself being a product of the natural radioactive decay chain of uranium (U^{238}) which is found in soil and rocks around the world (Adams, 1964). Both radon and uranium emit gamma rays. Due to uranium's enormously long *half-life* of billions of years (half-life is a term that corresponds to the time period in which half of the atoms decay into another element, e.g. from uranium into radon), both of these radioactive substances will retain their presence at the same concentrations, thus the amount of radioactivity caused will remain the same (Roper, 1990). Exposure to excessive concentration of radon, only presents a health risk in low elevation indoor enclosures such as basements, and is known to increase the risk of developing lung cancer (National Cancer Institute, 2004). Radon and its floating radioactive products such as polonium (Po^{218}) and lead (Pb^{214}) can be absorbed through inhalation. Heavy metal particles therefore accumulate inside the body as radon decays. Along with other gases such as oxygen and carbon monoxide, radon readily dissolves in the blood and circulates throughout the body. So, radon is sucked in along with air whenever we breathe. It can also leave the body by exhalation through the lungs or sweating through skin. Its seriousness is reported by (Environment Protection Agency, 2003) that over 20 000 people die in the US alone because of radon induced lung cancer.

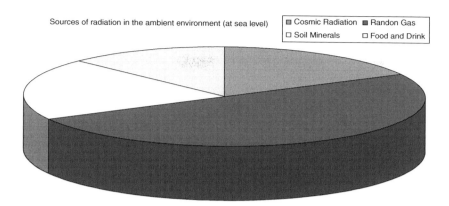

Sources of radiation in the ambient environment (at sea level)

☐ Cosmic Radiation ■ Randon Gas
☐ Soil Minerals   ☐ Food and Drink

Figure 8.19   Radiation sources

An average person receives a higher radiation dose from radon at home than from anywhere else with other natural and manmade radiation sources combined. From the composition shown in Figure 8.19, we can see that about half of all the natural radiation sources come from radon.

In the natural world, any inhaled radon atom that decays before it has a chance to leave the body will form heavy metal particles that accumulate in the lungs and tracheobronchial tree, predominantly in bifurcations. Subsequent radioactive decay of the accumulated heavy metal may emit sufficient energy to damage surrounding epithelial cells. If trapped in the blood stream, there is also a small risk of causing leukemia or sickle cell anemia due to radioactive residues left by radon decay.

Cosmic radiation originating from outer space and the sun, consists of energetic charged particles such as protons and helium ions, is known to affect air travellers. The biological damage caused by subatomic particles is widely believed to be more serious compared to X-rays or Gamma rays. The intensity of cosmic radiation depends on altitude, latitude, and solar activities (World Health Organization, 2005). At a cruising altitude of around 33 000 ft (10 000 m), an airplane is subject to cosmic radiation of some 100 times more than that at sea level. The cosmic radiation intensity generally increases as we fly away from the equator towards the poles because of the diminishing shielding effect of the earth's magnetic field. On average, a few hundred flying hours per year would absorb a comparable amount of radiation dosage by an average person on the ground (Lewis, 1999).

Energy emitted by radiation, both X-ray and radioactivity, can carry sufficient energy to trigger genomic changes to the cell's DNA structure, including mutation and transformation. The consequential effect of genetic mutations and chromosome aberrations may cause birth defects to future generations if the defective gene is carried. Another potential problem is chemical radicals that can be created inside the cell.

Radiation risk to the foetus is higher than to children as the excessive energy can damage fragile embryonic cells. Children are more susceptible to radioactive emissions due to the combined effect of their rapidly dividing cells and higher breathing rates, the latter translates to breathing in more radioactive radon gas. A single X-ray dose to a pregnant woman in the first six weeks of pregnancy can lead to as much as a 50% increase in cancer and leukemia risks to the unborn baby. Carcinogens cause random damage to the chromosomes and DNA

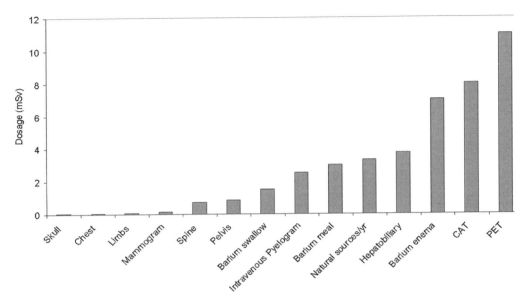

Figure 8.20   X-ray dosage

molecules inside the cell nucleus. While such damage normally destroys the cell completely, there is still a risk that a partially damaged cell can survive and reproduce with its defects sustained. Such a cell can then proliferate in a cancerous behaviour that ultimately develops into a cancer tumour.

So, how much is 'too much'? Quantitatively describing the amount of radiation (from medical diagnosis and the natural environment alike) can sometimes be confusing because different standards and units exist. A *millirem, mrem, millirad* and *mrad* are all identical measurement units. Also, *1 mSv* is equivalent to *100 millirem*. To understand how much one unit of mSv is, we generate a chart that shows the typical dose of X-ray based on figures given by (Wall, 1997) and (UNSCEAR, 2000); these are summarized in Figure 8.20. This chart shows us that the amount of dosages from a few X-ray examinations combined would be very insignificant compared to what an average person is subjected to from natural radiation annually. Cumulative exposure from CAT scans may slightly increase the risk of cancer (Reinberg, 2009).

References

Adams, J. A. S., Louder, W. M., Phair, G., and Gottfried, D. (1964), *The Natural Radiation Environment*, US Department of Energy Nuclear Testing Archive, Accession Number: NV0053511, Document Number: 57452.

Als-Nielsen, J. and McMorrow, D. (2001), *Elements of Modern X-ray Physics*, John Wiley and Sons, ISBN: 0471498580.

Bloch, C. (2009), Federal Telemedicine News: News briefs and information from Federal agencies and Capitol Hill on government activities, legislation, and grants of interest to the telemedicine, telehealth, and health IT community, 8th September, http://telemedicinenews.blogspot.com/2009/09/nih-awards-grants.html

Briggs, D. (2003), Environmental pollution and the global burden of disease, *British Medical Bulletin*, 68(1):1–24.

Brinker, M. R., Garcia, R., Barrack, R. L., Timon, S., Guinn, S., and Fong, B. (1999), An analysis of sports knee evaluation instruments, *American Journal of Knee Surgery*, 12(1):15–24.

Brunet, M. E., Brinker, M. R., Cook, S. D., Christakis, P., Fong, B., Patron, L., O'Connor, D. P. (2003), Patellar tracking during simulated quadriceps contraction, *Clinical Orthopaedics and Related Research*, 414:266–275.

Cartwright, F. F. (2004), *Disease and History*, 2/e, Sutton Publishing, ISBN: 075093526X.

Centers for Disease Control and Prevention (2009), Outbreak of Swine-Origin Influenza A (H1N1) Virus Infection – Mexico, March–April 2009, CRC, USA, 30 April, http://www.cdc.gov/mmwr/preview/mmwrhtml/mm58d0430a2.htm

Chow, D. H. et al. (2006), The effect of backpack weight on the standing posture and balance of schoolgirls with adolescent idiopathic scoliosis and normal controls, *Gait Posture*. 2006 Oct, 24(2):173–181.

Environment Protection Agency (2003), *Assessment of Risks from Radon in Homes* (EPA 402-R-03-003); A Citizen's Guide to Radon: The Guide to Protecting Yourself and Your Family from Radon, US Environmental Protection Agency (EPA 402/K-09/001), January 2009. http://www.epa.gov/radon/pdfs/citizensguide.pdf

Chow, D. H. K., Leung, K. T. Y., and Holmes, A. D. (2007), The effects of load carriage and bracing on the balance of schoolgirls with adolescent idiopathic scoliosis, *European Spine Journal*, 16(9):1351–1358.

Cleverley, M. (2009), How ICT advances might help developing nations, *Communications of the ACM*, 52(9):30–32.

Darrow, A. A., Johnson, C. M., Ghetti, C. M. and Achey, C. A. (2001), An analysis of music therapy student practicum behaviors and their relationship to clinical effectiveness: An exploratory investigation, *Journal of Music Therapy*, 38(4):307–320.

Deer, J. C. B. and Kemp, P. (2006), *The Oxford Companion to Ships and the Sea*, 2/e, Oxford University Press ISBN: 019920568X.

Denton, F. T. Gafni, A., and Spencer, B. G. (2002), Exploring the effects of population change on the costs of physician services, *Journal of Health Economics*, 21:781–803.

eHealth Europe (2009), *FASS up to 100,000 telecare deployments*, online report published on 9 July, 2009: http://www.ehealtheurope.net/News/4985/fass_up_to_100000_telecare_deployments

Filler, A. G. (2009), The History, Development and Impact of Computed Imaging in Neurological Diagnosis and Neurosurgery: CT, MRI, and DTI, *Nature Proceedings*, 13 July.

Gu, J., Barker, D., and Pecht, M., Health Monitoring and Prognostics of Electronics Subject to Vibration Load Conditions, *IEEE Sensors Journal*, 9(11):1479–1485.

Kantor, M. and Irving, L. (1997), *Telemedicine Report to Congress*, US Department of Commerce in conjunction with the Department of Health and Human Services, 31 January, http://www.ntia.doc.gov/reports/telemed/index.htm

Kelly, J. (2005), *The Great Mortality, An Intimate History of the Black Death, the Most Devastating Plague of All Time*, New York.

Lewis, B. J. et al. (1999), Cosmic radiation Exposure on Canadian-Based Commercial Airline Routes, *Radiation Protection Dosimetry*, Oxford University Press, 86(1):7–24.

Li, J. et al. (2006), Development of a Broadband Telehealth System for Critical Care: Process and Lessons Learned, *Telemedicine and e-Health*, 12(5):552–560.

Little, L. K. (2008), *Plague and the End of Antiquity: The Pandemic of 541–750*, Cambridge University Press, ISBN: 052171897X.

Massachusetts Institute of Technology (2008), Bacterial 'Battle For Survival' Leads To New Antibiotic. *ScienceDaily*, February 27, http://www.sciencedaily.com /releases/2008/02/080226115618.htm

Maugh, T. H. (2002), An Empire's Epidemic: Scientists Use DNA in Search for Answers to 6th Century Plague, *Los Angeles Times*, May 6.

National Cancer Institute (2004), *Radon and Cancer: Questions and Answers*, US National Institutes of Health Fact Sheet. http://www.cancer.gov/images/Documents/9f80b377-3962-4898-bd71-d1ebcee2d32d/fs3_52.pdf

National Statistics (2009), http://www.statistics.gov.uk/cci/nugget.asp?ID=949

Nobel Prize (2009), The Nobel Prize in Physiology or Medicine 2009: 'for the discovery of how chromosomes are protected by telomeres and the enzyme telomerase' http://nobelprize.org/nobel_prizes/medicine/laureates/2009/

Reinberg, S. (2009), As CT Radiation Accumulates, Cancer Risk May Rise, *US News and World Report*, 31 March, http://health.usnews.com/articles/health/healthday/2009/03/31/as-ct-radiation-accumulates-cancer-risk-may-rise.html

Roemer, M. I. (1993), *National Health Systems of the World: The issues*, Oxford University Press.

Roper, W. L. (1990), *Toxicological Profile for Radon*, Agency for Toxic Substances and Disease Registry, US Public Health Service in collaboration with the US Environmental Protection Agency. http://www.bvsde.paho.org/bvstox/i/fulltext/toxprofiles/radon.pdf

Santiago-Palma, J. and Payne, R. (2001), Palliative care and rehabilitation. *Cancer.* 924:1049–1052.

Scott, E. (2007), Music and Your Body: How Music Affects Us and Why Music Therapy Promotes Health, http://stress.about.com/od/tensiontamers/a/music_therapy.htm

Scott, T. (2009), Digital TV software provides talking menus for the visually impaired, deaf and elderly, *Accessibility News International*, 2 September.

Sheehan, D. K. and Forman, W. (2003), *Hospice and Palliative Care: Concepts and Practice*, 2/e, Jones & Bartlett Publishing, ISBN: 0763715662.

Shneiderman, B. (2005), *Designing the User Interface: Strategies for Effective. Human-Computer Interaction*, Addison-Wesley Publishers, Reading, MA

Standley, J. M. and Prickett, C. A. (1994), Research in Music Therapy: A Tradition of Excellence, *The National Association for Music Therapy, Inc.*, ISBN 0-935868-71-2.

UNSCEAR (2000), *Sources and effects of ionizing radiation, Vol. 1: Sources*, United Nations Scientific Committee on the Effects of Atomic Radiation, New York.

Vergados, D. D. (2007), Simulation and Modeling Bandwidth Control in Wireless Healthcare Information Systems, *Simulation*, 83(4):347–364.

Wall, B. F. and Hart, D. (1997), Revised radiation doses for typical x-ray examinations. *The British Journal of Radiology*, 70:437–439.

Willforth, J. C. (1985), Medical Radiation Protection: A Long View, *American Journal of Roentgenology*, 145(6):1114–1118.

World Health Organization (2005), Information Sheet: *Cosmic Radiation and Air Travel*, November, http://www.who.int/ionizing_radiation/env/cosmic/WHO_Info_Sheet_Cosmic_Radiation.pdf

Zhang, H., Small, M., and Fu, X., Staged progression model for epidemic spread in homogeneous and heterogeneous networks. *Journal of Systems Science and Complexity*, 22(4).

9

Future Trends in Healthcare Technology

Throughout the previous eight chapters, we have discussed how telemedicine and related technologies assist various aspects of healthcare and medical practices. Most of the technologies have a long proven history. Data communications evolved from the first telephone by Graham Bell and Elisha Gray formed the basis of many modern telemedicine systems deployed throughout the world today (Bashshur, 2009). Technology advances and innovative breakthroughs are opening a wide range of possibilities in medical and healthcare services. The Health Informatics Review Report published by the UK's (Department of Health, 2008) discusses the importance of delivering better and safer healthcare services through research, planning and management in health informatics.

Telemedicine is certainly an important core technology for healthcare service delivery. Evolving technologies make data communication faster, safer, and more economical. Reliability is the most important aspect of any system as we learned in section 5.4. Indeed, an unreliable system would be useless no matter what it is capable of doing. We shall begin this chapter by looking at how reliability can be optimized.

9.1 Prognostics in Telemedicine

The word 'prognostics' usually refers to a forecast of what might happen based on signs or symptoms in making a prognosis. This implies prognostics can predict what might happen to a system so that reliability can be assured. For example, we can deduce a schedule for when calibration or preventive maintenance has to be carried out before the system fails. The word 'Prognostics' is defined in by the Centre for Prognostics and System Health Management as:

'Prognostics' is an engineering discipline focused on predicting the time at which a component will no longer perform a particular function. Lack of performance is most often component failure. The predicted time becomes then the 'remaining useful life' (RUL). The science of prognostics is based on the analysis of failure modes, detection of early signs of wear and aging, and fault conditions. These signs are then correlated with a damage propagation model. Potential uses for prognostics

Telemedicine Technologies: Information Technologies in Medicine and Telehealth   Bernard Fong, A.C.M. Fong, and C.K. Li
© 2011 John Wiley & Sons, Ltd

is in condition-based maintenance. The discipline that links studies of failure mechanisms to system lifecycle management is often referred to as 'prognostics and health management' (PHM), sometimes also 'system health management' (SHM)...

Note, incidentally, that the word 'health' here refers to the system's health status rather than human health as in this text. In essence, what we would like to accomplish is by deploying prognostics and health management (PHM) techniques to optimize the health of medical systems so that these systems can in turn optimize human health. Based on this definition, PHM can be used for condition based maintenance for any system taking into consideration any performance degradation during its operational life. Indeed, PHM has been a proven technology widely used in many consumer electronics products (Vichare, 2006). Of course, medical devices are made up of electronic components, the main difference between those for consumer electronics and medical systems are mainly reliability and precision in terms of requirements since the impact of a failure would be far less on the former than the latter. PHM ensures reliability on electronic components and devices, electronics packaging, product reliability and systems risk assessment (Pecht, 2005). Proper prognostic health management can ensure hardware reliability.

Network outage, usually the main cause of telemedicine system failure, refers to the problem where the wireless link is temporarily disrupted, which may be due to intentional activity such as system maintenance or upgrade. The weakest link of the entire telemedicine system lies with the network transport section which, depending on the type of wireless network used, can span from several kilometres within a city to across continents. As discussed earlier in section 2.4, a number of factors can cause severe disruption along the signal propagating path.

Network breakdown is usually due to stochastic link failures (Egeland, 2009), where statistical modelling can describe its occurrence due to certain events. Prognostics techniques will require information about network data traffic to be collected and analyzed in order to ensure maximum reliability and availability. It uses data transmission performance of the wireless network to detect potential and future problems. In wireless telemedicine systems, most problems are caused by either wireless link or hardware failure. Prognostics enables link outage prediction through statistical modelling as well as the maintainence of optimal balance between reliability and performance. With condition-based monitoring, the network health can be maintained by adjusting a number of parameters in response to any performance degradation. For example, adaptive power control and the data throughput can be dynamically adjusted according to network condition. Prognostics may also entail the use of different modulation schemes. Although QPSK is very robust with a relatively long range offered when compared to higher order modulations, more spectrum may be necessary, particularly in drier areas where less rain is recorded.

Rain is usually the most influential factor in the reliability of outdoor wireless communications. As such, adequate link margin must be allocated to combat the effects of rain-induced attenuation (Fong, 2003). Selection of an appropriate carrier frequency, primarily determined by licensing, will provide a trade off between bandwidth and range. Generally, frequencies of no more than 10 GHz will be much less affected by rainfall while having a channel of narrower bandwidth. Hub placement is also an important consideration to ensure maximum network reliability, infrastructure cost and coverage will decrease with increased hub spacing thus there is an economic trade off. This also leads to an issue of selecting the optimal point-to-multipoint (PMP) antenna patterns (Viikari, 2007). Condition-based network

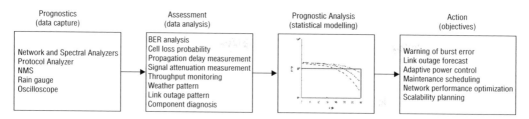

Figure 9.1   Prognostics framework

monitoring also allows control of sector to sector interference with frequency diversity and spatial diversity enabling high frequency reuse. This would eliminate the requirement for media access control (MAC) that would save overhead for improved bandwidth efficiency. The statistical information obtained can be used for computing the adequate margin to ensure network reliability irrespective of any change in operating environment.

Having looked at how PHM can monitor the condition of various parts of a telemedicine system, let us take a look at how PHM can be implemented. PHM relies on computational modeling of a known data set (Pecht, 2009). Relevant data can be collected during normal operation of the telemedicine system. For example, information about the data in transit can be used to construct a statistical model that describes the network status. Any abnormally long packet delay or excessive data packet loss may indicate a network congestion or node failure. This kind of problem can be easily diagnosed by PHM techniques. In some systems, PHM can be implemented with diagnostic built-in-test circuitry installed. Other implementation options include software-firmware systems for fault identification and isolation that incorporate error detection and correction functionalities, self-checking and self-verification circuits. These circuits can be small pre-calibrated cells that fit into small biosensors. The task that they all share in common is to collect operational data to monitor any performance degradation. In addition to operational reliability, PHM models and tools can also optimize maintenance planning and assessing Return-of-Investment (ROI). Figure 9.1 summarizes the process of PHM implementation.

Statistics about the network's 'health' (its condition) is usually collected from a Network Management System (NMS). The NMS is usually a piece of software installed on a computer that monitors the network condition and predicts a network outage when performance deteriorates. Figure 9.2 shows a scenario where a link failure can be expected when the rain becomes heavier. Heavier rain causes more signal attenuation hence reduces link availability. The link condition is continuously monitored based on information about data transfer so that certain network parameters can be adjusted in order to ensure data transmission reliability as the network condition degrades. Some networks do not have direct links between the transmitter and receiver so that data transmission must go through some nodes or repeaters. When the network degrades, certain paths along the network may be temporarily disconnected from the overall network to avoid network disruption. When a node fails, as shown in Figure 9.3 where part of a network with multiple nodes is depicted, each data packet can travel through any path along a combination of nodes across the network. When a link within the network fails, data packets can be re-routed based on known information about the network condition and the location of the failed node. Packets that experience abnormal delay or loss and have gone through a certain route would indicate that the route concerned is no longer reliable and hence

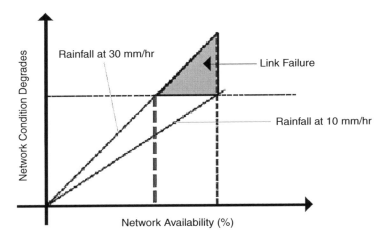

Figure 9.2   Network failure model

no more packets should be routed through that part of the network. The lost packets may need to be re-transmitted through other routes.

Data-driven prognostics techniques monitor network health through analysis of various network parameters, these include data loss, packet delay, latency, BER and E_b/N_o that tell us how well the network is performing. NMS or protocol analyzer, usually a piece of software package that is installed on a network computer console, provides such information about the health state of the entire network. Typically, an NMS or protocol analyzer will generate a list of information related to packets that travel across the network. Many NMS also proactively detect abnormity such as that shown in Figure 9.3 with a link outage somewhere across the network. Data packets can be automatically diverted to the bottom path that does not exhibit any known problem.

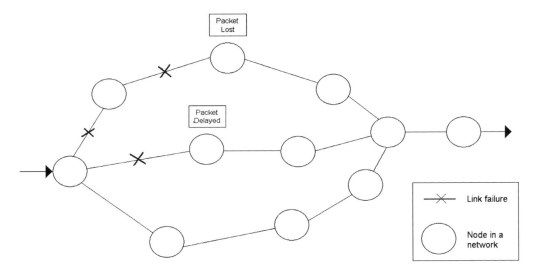

Figure 9.3   Network breakdown with re-routing

Fault detection is often carried out when certain network parameters fall below a certain pre-determined threshold level. Further diagnostics can attempt to fix a problem depending on the nature of problem. For example, more system fade margin can be assigned to areas where heavy rain can severely affect a wireless link.

9.2 The Aging Population: Home Care for the Elderly

The retirement of baby boomers born after World War II will certainly become a more serious problem in most developed countries (Bloom, 2004). The combination of declining fertility and increased life expectancy will lead to a substantial increase in the number of senior citizens. As we have looked at telecare technologies for assistive elderly care in section 8.2, we shall go through a case study on what lies ahead by taking a look at elderly home care in the UK. To assist with daily activities, we look at the TV-based telemedicine system '*Nexus TV*®', developed by Ocean Blue Software for elderly care.

9.2.1 *TV-Based Assistive Edutainment Monitoring: A Case Study*

This solution is capable of delivering healthcare savings of around £5 billion each year to the NHS, other healthcare providers and families by providing an alternative to residential care that allows the vulnerable to retain their independence. Medical benefits include automatic alerts reminding users which medication to take, and when. They will be able to order repeat prescriptions by scanning a barcode with the remote control and sending it to their dispensing chemist. It will also be possible to send photographs to clinicians for advice, and there will be direct links to doctors' surgeries and NHS Direct.

As shown in the system diagram of Figure 9.4, the backbone technology is a TV set top box that provides assistive support, companionship, independence and security for senior citizens. In addition to standard digital TV features, it provides users with talking menus that facilitate users who are visually impaired. It can also be linked to the Internet and call centre for a range of monitoring and support. Health monitoring and security can also be accessed through the TV screen. Our case study essentially uses a customized TV set top box that serves as a communications hub that links elderly users to different health monitoring devices and service centres together.

Cameras can be connected to Nexus TV to screen callers, and a local social networking service, based on an interactive message board, will enable people to stay in touch with those around them. Entertainment services will include *Freeview* digital TV, talking TV guides and menus, and downloadable audio books. The technology will also support the development of third party software applications, opening the door to a wealth of additional entertainment, games, education, and other services.

At this point, let us take a brief look at some background on Freeview digital TV. Hardware & Software specifications for UK Digital TV receivers have been set by the DTG (Digital Television Group). These DTG specifications form the minimum requirements and help create benchmarks in an emerging marketplace. These specifications should be adopted by consumers to prevent non-standard product entering the UK marketplace and also helping to reduce product return rates for Freeview digital TV receiver devices. The set top box architecture is shown in Figure 9.5. MHEG-5, the UK DTG interactive media standard for sending and receiving digital media objects, is a required software standard for Freeview products.

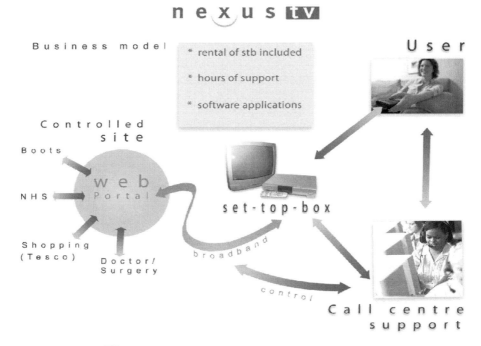

Figure 9.4   Nexus TV™ developed by Ocean Blue Software. Reproduced with permission of © 2009 Ocean Blue Software Ltd

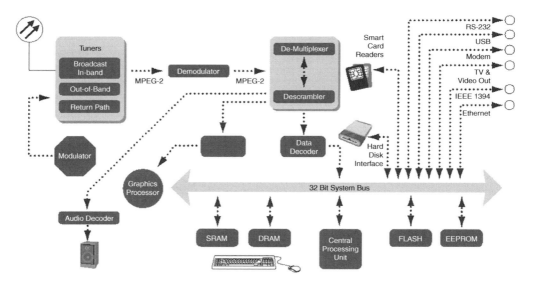

Figure 9.5   Architecture of an elderly assistive care TV set top box

The MHEG-5 return path is an extension to MHEG-5 that gives more bandwidth to broadcasters by using the Internet. There are two main reasons behind this extension. For services such as the BBC a return path in MHEG-5 allows them to add streaming and static content on top of broadcast content, thus allowing for much more data to be available via MHEG-5. For instance, a menu option on MHEG-5 saying 'Football Highlights', that when selected would go to a URL that would have been specified in the MHEG-5 application and obtain the stream and any additional MHEG-5 content via the return path. The other proposed use for the extension is to allow commercial 'retail' broadcasters to allow people to purchase product via the return path. The viewer would be able to select a button or option to purchase, which would result in a URL (that is part of the MHEG-5 application) being accessed and a secure transaction will take place. For this model the purchaser would have to be pre-registered with the retailer meaning the transaction would not involve the transmission of sensitive data such as credit card numbers.

The MHEG-5 return path is not being considered as a true IPTV solution, users will not be able to enter their own URL information or access services outside those that were pre-set by the broadcasters. The IP connection is to remain hidden and embedded in the MHEG-5 application. For example, if a new broadcaster was to come on the scene to provide Video on Demand service, it could be used to select video content from that broadcaster's vault, but again only through MHEG-5 interaction.

There has been industry wide recognition for the need to enhance interactive services through the introduction of a return path. A return path will allow content providers on digital terrestrial to offer viewers a new range of applications such as networking between elderly residents with interactive games, chat-like services allowing viewers to send in comments and opinion, video on demand and interactive transactions for purchasing goods and services.

Access to local services of all kinds, from taxis to food stores, will be available using a dedicated database, accessed via the television. Shopping for home delivery will be easily accessible too, optionally making use of barcodes scanned with the remote control. In terms of data communications, it uses a broadband connection just as those used for home Internet access. Links to entities such as NHS Direct, Social Services, Local Councils and even placing an order with a local supermarket are all done with the broadband link.

The system that complies with DTG standards need to be tested for a number of parameters including:

- MHEG Application Programming Interface testing
- Common Interface testing
- Service Information and (PSI) Programme Service Information signalling operations, including Electronic Programme Guide
- Audio and video testing, including Active Format Descriptors which tell the Digital TV receiver which parts of the picture are important
- Subtitle and audio description stream testing
- RF (Radio frequency) performance testing which tests the performance of the set top box

9.2.2   Smart Home Assistive Technologies

The idea of context home environmental interventions and assistive technology devices for elderly independence was raised a decade ago by (Mann, 1999). The idea of assistive home automation has been proposed with smart home control (Grossi, 2008). Smart home technology is a collective term for assistive communication technologies as used within a home, where the various components communicate via a wireless local area network (WLAN) as well as the Internet. Smart home technology supports a variety of electronics and appliances to communicate with each other and perform a variety of tasks. Let us illustrate this by taking a look at what a refrigerator can do. First, it can suggest what to drink based on what is stored inside it, the ambient temperature and time of the day. An Internet-enabled refrigerator can download recipes based on what is stored inside. It can communicate with a microwave oven to prepare the cooking power and time for the given mix of ingredients (Barthold, 2002). Smart refrigerators are reported to improve health together with multimedia technologies (Luo, 2008). Kuwik (2005) has also illustrated the potential of a smart medical refrigerator for senile dementia sufferers to deal with prescribed drugs and to monitor the use of insulin by diabetes sufferers. A refrigerator can be programmed to keep an inventory of items stored, track how long an item has been stored there and whether anything has expired, as well as when something needs to be replenished.

In addition to the refrigerator, smart home technology can be implemented in virtually all kinds of home appliances for more automation and intelligence. When used in conjunction with a home network or the Internet, different devices can communicate with each other. They can even facilitate communication between different users, caregivers, and device manufacturers. Smart home technology has already been widely implemented in many kitchens (Kranz, 2007) and entertainment in living rooms (Palazzi, 2009) for various control functions. Integrating smart home technology with telemedicine in an elderly patient's home, a range of possibilities can be offered in addition to the cooking and entertainment functions outlined above. (Demiris, 2004) assessed the use of smart home technology for preventing or detecting falls, assisting with visual or hearing impairments, improving mobility, reducing isolation, managing medications, and monitoring physiological parameters; while he reported that the main concerns users expressed were user-friendliness of the devices, lack of human response and the need for training tailored to older learners. (Rialle, 2004) also reported that the large diversity of needs in a home-based patient population requires complex technology. Such need demands data acquisition and wireless communication technology that elderly users with minimal training would feel comfortable using at home.

Artificial and computational intelligence plays an important role in providing assistive technology to elderly people. The diverse range of activities is mainly supported by these three major domains:

Communications to the outside world: between people and devices

Video conferencing can be set up between friends and family members for networking and with caregivers for advice and assistance. For example, a TV set with a small webcam can facilitate real-time communication between different parties. Without keyboard or mouse, a user can get connected with a remote control or speech command. They can even participate in a range of video games with people far away. Around the elderly user, doctors and other caregivers are well within reach. All these are made possible by telecommunications.

Users are connected to devices for virtually countless numbers of activities ranging from personal comfort to critical care. Devices are also interconnected so that comprehensive services can be supported. For example, we mentioned that a refrigerator can be connected to a microwave oven and recipes can be downloaded from the Internet. So, based on what is stored in the refrigerator cooking instructions can be provided to the user. Other appliances such as food processors and coffee makers can also be linked together to provide assistance for preparing a complete meal with ease.

Communication technology in a smart home environment not only benefits elderly residents but also their relatives. They can be assured that they are kept informed of the elderly user's well-being by getting far more information than what can be provided by a mobile phone. When one is thousands of miles away from an elderly relative or working just a few blocks away, they can be assured that an alert will be received in case of an emergency and help is always available to them.

Sensing the surrounding environment: on and around the user

Quality of life enhancement by monitoring activity of the user and what is around can be best supported by telemedicine and related technologies. Monitoring devices such as accelerometers, pressure sensors, motion detectors and video cameras can be either discretely or collectively installed in the smart home to collect details about the status of an elderly person. Sensors are used in areas from logging when the door has been opened to tracking the movement of a user. As discussed throughout the text, there are different sensors for health monitoring. Computational intelligence can also collect user data to learn and analyze data from long-term patterns of user behaviour. This can serve many objectives including rehabilitation progress, warning of abnormality, and active prevention of a fall.

For those who have cognitive or visual impairments, sensors can also help deal with the shortcomings, users can be reminded of undertaking daily activities such as switching off gas stoves and taking medication. They can be warned of any forthcoming hazards like walking towards a staircase or slippery surface. Smart home technology can provide contextual guidance and warnings in hazardous situations according to environmental conditions so that preventive measures can be taken.

Used in conjunction with a telemedicine network, a doctor can retrieve information about the user and view up-to-date electronic patient record. Whether the user has been eating or drinking properly as well as other behavioural variations can be easily observed.

Emotional intelligence: remaining happy and healthy

The importance of remaining happy to live life to the fullest is well-shared by those who have retired. After all, the vast majority of elderly persons have contributed decades of hard work to the community in different capacities. Quantitative assessment of how well a smart home system performs can be easily measured for the communication network and sensor network. There are parameters such as bit error rate (BER), latency, data loss, and indeed everything that we have covered in Chapter 2. What about happiness, self-confidence and self-esteem? What we have discussed so far only takes care of the user's physical well-being. What about technology that deals with emotional issues like loneliness and fear?

For those who live alone, a 'talking machine' can help, a dummy that initiates a conversation when someone approaches, presents a news brief about what is happening around the world and suggests going out for a meal. Body language and habitual behaviour can provide information about the user's psychological well-being. With speech recognition, social interactions can be made possible (Chen, 2007). Emotional intelligence applications can take actions according to the user's mood, when the user is bored the system can suggest some entertainment. For a talking machine, the system can adjust the tone of conversation according to the user's mood detected.

Smart home assistive technologies also help with energy saving by regulating temperature and illumination, air conditioners and blinds can be adaptively controlled by sensors installed throughout the home. Likewise, lights can be automatically turned on when a user is in the room and the ambient light from the windows fall below a certain level. Further, medication dispensing can be connected to the system and automatically locks itself when not needed (Cheek, 2005). As with a refrigerator, for those who require long term medications each drug can also be automatically tracked and order new stock before they run out. Smart home technology gives a totally different flexibility and functionality than traditional home networks.

9.3   Clothing Technology and Healthcare

As with smart home technology, artificial and computational intelligence can also be embedded in clothing for various tasks ranging from lost person tracking to professional sports training (Mann, 1996). Smart clothing involves far more than self-heating and glowing textiles (Gould, 2003). Traditionally, smart clothing deployed in specific areas like space suits used by astronauts is dense with miniaturized electronics but it is increasingly becoming possible for telemedicine applications as electronics and components become smaller, cheaper, and more structurally flexible. Smart clothing technology can trace a lost person by embedding an RFID transponder chip (Hum, 2001). Similar to that used in airports and postal systems for item tracking, RFID tags are used for elderly people who are disoriented or having lapses of memory (Dunne, 2005). Such personal identification can also serve as an alternative to door locking so that keyless entry can be supported. Conversely, such technology can be used to restrict individual freedom. The locking of doors and similar measures can be used to control access in restricted areas or for child safety.

Smart clothing has been used for monitoring body fluids, in rehabilitation and chronic disease monitoring (BBC News, 2007). Functions as an active device, many with embedded electronics having the ability to store and manipulate data; display information, input data, and communicate to the outside world, all these assist with various activities for both patients and caregivers. Some can offer passive protection in much the same way as air bags in motor vehicles. For example, detecting the presence of hazardous chemicals in the air, rapidly deploying a protective filtering mask, changing colour according to the environment for camouflage, and projecting an image of the scene behind the wearer for perceived invisibility. Although most are powered by batteries just like ordinary wearable devices, some can actually generate power from the wearer's movement in much the same way as the winding mechanism of an automatic wristwatch as shown in Figure 9.6. Its operating principle is quite simple. The eccentric weight of the rotor that turns on a pivot caused by movements of the user's wrist causes the rotor to pivot back-and-forth on its shaft, which is attached to a ratcheted winding mechanism. The motion of the wearer's arm is thereby translated into the circular motion of the rotor that, through a series of gears, the mainspring is wound automatically by the natural motion of the wearer's wrist. Embedded electronics in clothes can be powered by such a mechanism so that they can operate once worn.

Figure 9.6   Automatic winding movement mechanism

Although the power supply may not be as bulky as a battery, electronic components are usually rigid and bulky which contradicts the fact that clothes are made as soft and light as possible. Wearing comfort therefore becomes an important design issue. Another important design consideration is whether the clothes are washable. Something that can be washed just like ordinary clothes made of fabric and possibly with plastic buttons. With these basic requirements understood, let us take a look at a case study with an 'intelligent wristband' that continually monitors the blood glucose level of a type-1 diabetes patient.

A fabric wristband that consists of a light source, a photo sensor, timer, and a Bluetooth transmitter is shown in Figure 9.7. The electronic components are embedded in the fabric wristband. The controller illustrated in Figure 9.8 drives the infrared light beam and photo-sensor pair that measures the blood sugar level, where a certain percentage of the infrared beam is absorbed by the blood depending on the sugar content. The amount of the beam that is reflected therefore represents the amount of sugar present. The controller is also responsible

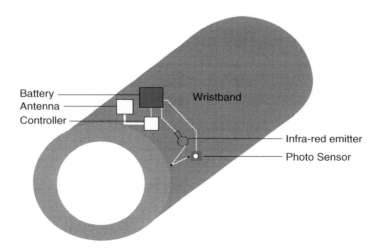

Figure 9.7   Glucose measuring wristband

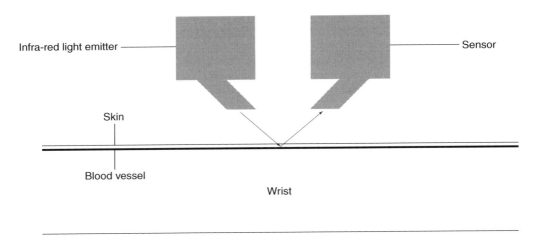

Infra-red light emitter

Sensor

Skin

Blood vessel

Wrist

Figure 9.8   Non-invasive optical glucose measurement

for capturing and forwarding the reading through a wireless link with a patch antenna. The captured data is stored-and-forwarded to a Java-enabled mobile phone via a Bluetooth link. The mobile phone serves as a console for storing and analyzing the measured data. It can also be connected to the user's family doctor where appropriate. Any abnormality detected can therefore alert the physician to any necessary follow-up action.

Calibration prior to use is an essential step to ensure reliability. This involves testing on a subject with known blood glucose level. A blood glucose laboratory test will be performed for the purpose of setting a reference value to calibrate the reading obtained from measuring the amount of infrared light absorption, since the total amount of light absorption also accounts for skin and tissue absorption. This will normally be accomplished by fasting blood glucose (FBG) level measurement prior to the first use. While calibration guarantees measurement accuracy for a certain period of usage time thereafter, prognostics and system health management can be effective in deducing the deviation from expected precision and any impact on measurement due to changes in environmental parameters, such as ambient temperature, humidity, shock, and skin condition.

9.4   Haptic Sensing for Practitioners

Haptic is a tactile feedback technology through touch. Haptic sensing reacts to the user's hand movement including forces, vibrations, and motions. This provides a user interface that utilizes the sense of touch. Note, incidentally, that tactile sensors that sense the amount of force exerted on the interface in a *somatosensory* system, of the peripheral nervous system (PNS) and the central nervous system (CNS), is not considered a haptic sensor. A somatosensory system is one that consists of receptors which respond to different stimuli and processing centres that generate sensory modalities. Control based on haptic sensing would be limited by fiction, precision, and lack of stimulus for the sense of touch (Smith, 1997). To understand more about haptics, we look at a control glove in Figure 9.9, where a number of sensors are found around the palm; these sensors are driven by real-time algorithms that interpret the hand's movement and drive an actuator controller. Here, the haptic mechanism conveys forces from the user's hand to the remote actuators. On the remote side, there are actuators and control circuitry that act upon the user's hand movement. What needs to be considered includes actuator size, precision, resolution, frequency, latency requirements, power consumption and cost of operation. The controller can be either closed-loop or open-loop. In closed-loop control, the controller reads sensor movement from the received signal, and then computes and executes the haptic output forces in real time based on the sensor movement. In open-loop control, a triggering event will activate the controller to compute and relate the haptic output signal to the actuator in real time.

One obvious application of haptics in telemedicine is remote robotic surgery (Okamura, 2004). One major advantage of using haptics for robotic surgery is for medical schools when students can practice on simulators so that there will be no risk of injuring a patient while learning to operate (Shen, 2008). Another important application is operations where visualization is not possible. The amount of force being exerted on an organ or tissue can be very delicately controlled and regulated by actuators. With robotic tele-surgery set up, a patient can be prepared by local hospital staff and operated upon by expert surgeons who can perform anywhere without travelling (Davies, 2000).

Protecting veterinary surgeons is also one major advantage of haptics in surgery. Dog bite injuries are risks that can be eradicated if the surgeon does not make direct contact with the dog being operated on (Overall, 2001). In fact, it is even possible to operate on an animal while it is kept in a cage with a robot inside. Surgeons can easily perform the operation outside the cage with tele-surgery.

As shown in Figure 9.9, any system that supports robotic tele-surgery requires a communication link that links the surgeon's hands to remote actuators, or a simulator in case of

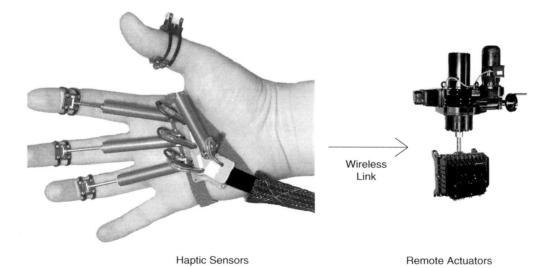

Haptic Sensors   Remote Actuators

Figure 9.9   Haptic glove

surgical practicing. This link needs to deliver the information that replicates exactly the hand's movement in real-time. In addition to control information for the actuators, a camera that acts as the surgeon's 'remote eye' also needs to transmit crystal clear images in real time back to the surgeon (English, 2005). The reliability and bandwidth requirements must be addressed. Here, we need to remember that video compression cannot be relied on because any loss in image detail can cause major disaster to an operation. The challenges of optimizing bandwidth efficiency and reliability still need to be thoroughly addressed before haptic control can be widely utilized in robotic surgery.

9.5   The Future of Telemedicine and Information Technology for Everyone: From Newborn to Becoming a Medical Professional all the Way Through to Retirement

One of the major advantages of wireless telemedicine is to provide medical services with a high degree of mobility. Advances in wireless communications have made new services possible over recent years. Mobile monitoring yields significant cost savings including patients who can be discharged soon after surgery for home recovery so as to minimize the duration of hospital stay without any adverse effects utilizing existing wireless home network for monitoring. Also, continual health monitoring can reduce demands for healthcare resources by maintaining optimal health. Other benefits include reduction in health insurance claims and loss in productivity. However, there are different levels of risks incurred when a patient is discharged from hospital early, depending on the nature of illness and physical state. Some may be taken care of by family members while others may require medical attention. This is because a vast range of possibilities exist in relation to different scenarios. For example, the risks associated with a patient after a coronary artery bypass will be very different from those of an acute myocardial infarction patient even though both have undergone cardiac operations.

Telemedicine technology offers many possibilities as various parameters can be monitored based on different circumstances; for example, throughout the text we have covered applications including: posture sensing for spinal injury or back pain, where a patient can enjoy continual health monitoring system that utilize accelerometer attachment or video imaging. These technologies can be used in situations such as post-surgery rehabilitation, prevention of effects of backpacks on children, and design for consumer devices such as massage chairs and optimal positioning of baby monitoring systems. Movement detection for knee and foot recovery, such as after an ACL (anterior cruciate ligament) operation assist with remote re-covery monitoring, a monitoring mechanism for walking or jogging to record parameters such as pace, distance covered, heart rate, and calories burnt. Wireless ECG measurement imposes far less movement restriction and reduces risk of infection due to pathogens on ECG telemetry lead wires.

Other services include alternative medicine addressing properties of different herbal medicines, and support of acupressure treatments. Such scalable informatics frameworks can also provide a better understanding of the genetic bases of complex diseases by analyzing vast amounts of data collected in genetic computation and patterns of disease spreading.

Telemedicine technology is truly something for all ages. In the above example, we have seen telemedicine applications that potentially everyone can utilize. For the remainder of this

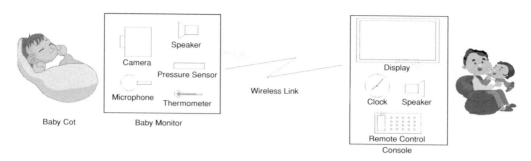

Figure 9.10 Wireless baby monitor

chapter, we shall walk through telemedicine with the little newborn baby girl Melody to see how technologies will assist her throughout all ages. We look at the possibilities that exist. Our aim is not to make wild forecasts on where technology is heading, but we would like to enlighten readers' thoughts on how current technologies can be pushed forward to assist with various tasks for people of all ages by recapitulating what the earlier text has covered.

When Melody leaves her mother's womb on the day of her birth, she is likely being attached with an RFID tag on the wrist. This is perhaps her first encounter of wireless communication technology. Since most babies look very similar to each other, RFID tag provides a safe and secure way to uniquely identify each newborn baby. Embedded to the tag is information including mother's name, date and time of birth, and any treatment provided as she is being monitored during the first few days of her life at hospital. Without understanding what is happening around her, Melody is already assisted by wireless communication technology that, although already in use elsewhere in very limited areas, was not available to her mother when born some three decades ago.

Melody's parents bring her home a few days after birth. There is a good chance that her parents have bought her a baby monitor, one that was described in section 7.3 as shown in Figure 9.10. With a video camera and sensors around Melody, her parents can leave her alone in the cot while enjoying some home entertainment in the adjacent room with the assurance that Melody is sleeping well. Pressure sensors ensure that the baby will not turn over while sleeping and alert her parents of any potential risk of rolling over. A microphone lets her parents hear what happens. With understanding about the sonic pattern of baby cries, speech processing algorithms also analyze Melody's cry and suggest the likely actions needed. For example, whether she wants attention or if she is hungry. Her parents can also look at how she is doing without going into the room. This also prevents disrupting Melody's sleep. Last but not least, the ambient environment is fully regulated with smart home technology.

When Melody becomes a toddler, she can enjoy regular medical check ups to ensure normal growth with self-diagnosis and testing performed at home. All data will be automatically captured and updated by linking to her personalized electronic patient record. Parameters such as BMI, blood glucose level, ECG will all be recorded while she undertakes normal activities. She can even see the doctor remotely through video conferencing.

Melody grows and eventually becomes a medical school student. Mobile learning (M-Learning) portals engage students to learn at anytime, and anywhere, and to encourage truly active learning and teaching. M-Learning refers to the use of mobile and handheld devices, such as PDAs, mobile phones, laptops and tablet PCs in facilitating teaching and learning in a much more convenient and efficient way to learn anytime, anywhere. M-Learning within other educational contexts is often faced by significant challenges in terms of technical support and infrastructure (Kukulska-Hulme, 2005). Problem-solving is recognized as the core competence by the statutory body and the stakeholders in local healthcare organizations (Lennox, 1998). New learners not exposed to typical problem solving environment may be neither aware of their roles required to address patients' problems nor ready to apply the knowledge or skills leading to a solution to meet patients' needs. However, the students may not realize that they are required to prove the mastery of intellectual and psychomotor skills within the context-based practices. The process of problem solving entails high-order thinking, in relation to skills in critical thinking and problem-solving. With the theoretical benefits learnt in classroom, physicians might adopt the cognitive procedures leading to possible solutions. The attributes and competence of problem-solving to provide safe practice are characterized with vigilance to individual and contextual issues, risk identification and management, error reduction, and search of practical solutions. In fact, clinical problem solving remains a mix of conceptual understanding and cognitive skills. Students may not aware of the fact that the performance has to do with a range of integrative attribute and skills that involve the abilities to integrate and synthesis of factual information, theoretical concepts or procedures. In response to the impetus of developing the clinical problem solving among medical students in an acute care setting, a simulated clinical problem-solving (SCPS) component will be developed in a mobile delivery format. Mobile learning platform provides a SCPS component which is structured as a self-study element that requires initiative and active participation. Learners are encouraged to drill the relevant skills in the SCPS with the integral pedagogy on creative thinking, self-directed learning and experiential learning. Also, Melody can get as many chances to practice her surgical skills as she wishes with the aid of haptic sensing surgical simulators with the knowledge that even if she makes a serious mistake no real damage will be caused. Telemedicine and related technology will certainly make learning a lot easier for future medical students.

As a physician, she can remotely track a patient's post-surgical recovery process with a health monitoring system based on existing home wireless network for analyzing data captured by medical devices, such as an oximeter temporarily installed at home, for transmission to the hospital. The range of post-hospitalization checks supported includes medication and nutritional administration, body temperature and SpO_2 readings. Such information can be analyzed and appended to the patient's medical record. Such a system can help reduce demands on hospital resources as well as travel time for patients and caregivers, particularly helpful for the elderly and disabled patients.

Utilizing consumer healthcare technology and network sensors for general health assessment for elderly and vulnerable patients is another major area that assists with a physician's duty. Telemedicine can integrate IT and the Bedside. A scalable informatics framework that will bridge clinical research data and the vast databases arising from basic science research in order to better monitor people's health. A home healthcare system is based on an existing home IEEE 802.11 WLAN that also facilitates simultaneous, independent connections between various networking devices such as computers and audio/video systems. Biosensors can also collect physiological data of a patient just like the above example. The system will offer flexibility to support a wide range of healthcare monitoring services. Such patient monitoring system is integrated into an existing home wireless LAN system that provides a number of network access and device control functions. This ensures minimal intervention is necessary within the patient's home throughout the monitoring process. This home network effectively provides an access point to an external telemedicine system that has a direct connection to the hospital. The external telemedicine system utilizes a telemedicine system that provides a direct connection to the hospital with an infrastructure. This can be accomplished by using a public network allowing hospital staff to perform remote diagnosis and consultation for patients without leaving home. Various instruments can be attached to the system depending on the type of data sought to monitor the patient's progress and response to any sudden change. Its set up is simple and economical with all equipment at the patient's home installed on a temporary basis. Flexible monitoring can be offered to patients by making use of wearable computers for capturing data from the biosensors so that the patient can move freely when being monitored.

Advances in telemedicine technology help Melody through her career into retirement. Routine activities of senior citizens can be supported by a multi-sensory telecare system as an electronic guard. Elderly people with special needs such as memory loss and cognitive impairment sufferers can be greatly benefited by technological advancements in HCI and wireless communications. A wearable therapeutic device provides general assistance, health monitoring, calling for emergency assistance, alerts and reminders, which can provide dementia sufferer with a peace of mind. Mobility is also an important consideration as the current system is primarily designed for users remaining at home. User-friendliness is an important design consideration since most senior citizens are not familiar with technology. Another major function is to collect information about users' health conditions, medication and nutritional intake, and fall history. Such clinical information will be analyzed on a regular basis for monitoring purpose. In addition, the clinical information can be connected to and shared with healthcare facilities, including general practitioners or hospitals using any wireless network. This feature is particularly suitable for older adults and cognitive impairment users who are recovering at home after hospitalization (such as after hip fracture surgery) while still under close surveillance by hospital staff. In addition, this feature can help reduce demands in hospital resources as well as travel time for elderly patients.

Alerts and reminders that assist routine activities such as medication intake, flush the toilet after use, safe use of gas stove and fire risk all make use of telemedicine for safety enchancement. The system can be designed to help an elderly person with various tasks and daily routine activities with attachments of appropriate instruments and biosensors. Medication reminders and instructions are automatically generated by embedding drug info on an RFID chip in the bag.

Here, we have concluded the chapter by looking at how telemedicine and related technologies can assist a wide range of tasks for a person from birth to retirement and beyond. Technological advancements certainly bring countless exciting opportunities for medical science and healthcare that benefit both practitioners and patients.

References

Barthold, J. (2002), Cable rule ready to the roost, *Telephony*, 6 May.

Bashshur, R. L. and Shannon, G. W. (2009), *History of Telemedicine- Evolution, Context, and Transformation*, Mary Ann Liebert USA, ISBN: 1934854115.

BBC News (2007), Smart clothes to monitor health, 11 June 2007: http://news.bbc.co.uk/2/hi/health/6740325.stm

Bloom, D. E. and Canning, D. (2004), Global Demographic Change: Dimensions and Economic Significance, *NBER Working Paper* No. W10817, October 19.

Cheek, P. (2005), Aging well with smart technology, *Nursing Administration Quarterly*, 29(4):329–338.

Chen, D., Yang, J., Malkin, R., and Wactlar, H. D. (2007), Detecting social interactions of the elderly in a nursing home environment, *ACM Transactions on Multimedia Computing, Communications, and Applications*, 3(1): 1–22.

Davies, B. (2000), A Review of Robotics in Surgery, *Proceedings of the Institution of Mechanical Engineers – Part H – Journal of Engineering in Medicine*, 214(1):128–140.

Demiris, G. et al. (2004), Older adults' attitudes towards and perceptions of 'smart home' technologies: a pilot study, *Informatics for Health and Social Care*, 29(2):87–94.

Dunne, L. E., Ashdown, S. P. and Smyth, B. (2005), Embedded clothing technology, *Journal of Textile and Apparel Technology and Management*, 4(3):1–11.

Egeland, G. and Engelstad, P. (2009), The availability and reliability of wireless multi-hop networks with stochastic link failures, *IEEE Journal on Selected Areas in Communications*, 27(7):1132–1146.

English, J., Chang, C. Y., Tardella, N., and Hu, J. (2005), A vision-based surgical tool tracking approach for untethered surgery simulation and training, *Medicine Meets Virtual Reality 13: The Magical Next Becomes the Medical Now*, IOS Press, pp. 126–132, ISBN: 9781586034986.

Fong, B., Rapajic, P. B., Fong, A. C. M. and Hong, G. Y. (2003), Polarization of received signals for wideband wireless communications in a heavy rainfall region, *IEEE Communications Letters*, 7(1):13–14.

Gould, P. (2003), Textiles gain intelligence, *Materials Today*, 6(10):38–43.

Grossi, F., Bianchi, V., Matrella, G., De Munari, I. and Ciampolini, P. (2008), An Assistive Home Automation and Monitoring System, *International Conference on Consumer Electronics, 2008*. ICCE 2008. Digest of Technical Papers. 9–13 Jan.

Health Informatics Review Report (2008), Department of Health, UK: http://www.dh.gov.uk/prod_consum_dh/groups/dh_digitalassets/@dh/@en/documents/digitalasset/dh_086127.pdf

Hum, A. P. J. (2001), Fabric area network – a new wireless communications infrastructure to enable ubiquitous networking and sensing on intelligent clothing, *Computer Networks*, 35(4):391–399.

Kranz, M. et al. (2007), Sensing technologies and the player-middleware for context-awareness in kitchen environments, *Fourth International Conference on Networked Sensing Systems, 2007*. INSS '07, 6–8 June, pp. 179–186.

Kukulska-Hulme, A. and Traxler, J. (2005), *Mobile Learning: A Handbook for Educators and Trainers*, Routledge, ISBN: 041535739X.

Lennox, A. and Petersen, S. (1998), Development and evaluation of a community based, multiagency course for medical students: descriptive survey, *British Medical Journal*, 316:596–599.

Luo, S., Xia, H., Gao, Y., Jin, J. S., and Athauda, R. (2008), Smart Fridges with Multimedia Capability for Better Nutrition and Health, *International Symposium on Ubiquitous Multimedia Computing, 2008*. UMC '08. 13–15 Oct., pp. 39–44.

Mann, S. (1996), Smart clothing: the shift to wearable computing, *Communications of the ACM*, 39(8):23–24.

Mann, W. C., Ottenbacher, K. J., Fraas, L., Tomita, M., and Granger, C. V. (1999), Effectiveness of assistive technology and environmental interventions in maintaining independence and reducing home care costs for the frail elderly, *Archives of Family Medicine*, 8:210–217.

Okamura, A. M. (2004), Methods for haptic feedback in teleoperated robot-assisted surgery, *Industrial Robot*, 31(6):499–508.

Overall, K. L. and Love, M. (2001), Dog bites to humans—demography, epidemiology, injury, and risk, *Journal of the American Veterinary Medical Association*, 218(12):1923–1934.

Palazzi, C. E., Stievano, N. and Roccetti, M. (2009), A smart access point solution for heterogeneous flows, *International Conference on Ultra Modern Telecommunications & Workshops, 2009*. ICUMT '09. 12–14 Oct, pp. 1–7.

Pecht, M. (2005), Part Selection and Management, John Wiley and Sons USA, ISBN: 9780471723875.

Pecht M. (2009), *Product Reliability, Maintainability, and Supportability Handbook*, 2/e, CRC Press USA, ISBN: 0849398797.

Rialle, V., Duchene, F., Noury, N., Bajolle, L. and Demongeot, J. (2004), Health 'Smart' Home: Information Technology for Patients at Home, *Telemedicine Journal and e-Health*, 9 July 2004, 8(4): 395–409.

Shen, X. et al. (2008), Haptic-enabled telementoring surgery simulation, *IEEE Multimedia*, 15(1):64–76.

Smith, C. M. (1997), Human factors in haptic interfaces, *ACM Crossroads*, 3(3):14–16.

Vichare, N. M. and Pecht, M. G. (2006), Prognostics and health management of electronics, *IEEE Transactions on Components and Packaging Technologies*, 29(1):222–229.

Viikari, V., Kolmonen, V. M., Salo, J. and Raisanen, A. V. (2007), Antenna pattern correction technique based on an adaptive array algorithm, *IEEE Transactions on Antennas and Propagation*, 55(8):2194–2199.

Appendix: Key Features of Major Wireless Network Types

Since the Internet era began some two decades ago, communication networks have been expanded to cover virtually every part of the world, reaching remote areas by wireless links. Wireless communication technology evolved over a century ago from the Maxwell equations that describe the fundamentals of electricity and magnetism (Huray, 2009):

Electric Field (E):

$$\nabla \bullet \vec{D} = \rho_v$$
$$\nabla \times \vec{E} = -\frac{\partial \vec{B}}{\partial t}$$

Magnetic Field (H):

$$\nabla \bullet \vec{B} = 0$$
$$\nabla \times \vec{H} = \vec{J} + \frac{\partial \vec{D}}{\partial t}$$

With interrelationship:

$$\nabla \times \vec{E} = -\frac{\partial \vec{B}}{\partial t}$$

$$\nabla \times \vec{H} = \vec{J} + \frac{\partial \vec{D}}{\partial t}$$

Telemedicine Technologies: Information Technologies in Medicine and Telehealth   Bernard Fong, A.C.M. Fong, and C.K. Li
© 2011 John Wiley & Sons, Ltd

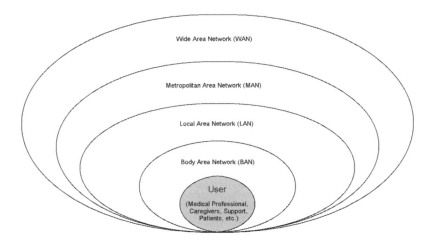

Figure A.1   Network classification based on coverage

We shall not go through the mathematical details but what we are interested in knowing is the fact that the above Figure A.1 shows what forms the basic blocks of wireless communications utilizing the relationships between electric and magnetic fields. Technical advances throughout the twentieth century have made many wireless transmission networks possible for an endless range of applications.

In the post-war era, television broadcast was probably one of the most popular wireless communication systems which is now used by billions of people throughout the world. While traditional broadcasting systems, radio and television services alike, use simplex point-to-multipoint (one way from one transmitter to many receivers) communication. It was around four decades ago when the cellular concept was proposed that has subsequently led to the development of many wireless communication networks today (Farley, 2007). Here, we look at some key features of major wireless network types and group them according to their intended coverage.

Body Area Network (BAN)

Technically known as IEEE 802.15 (http://www.ieee802.org/15/), it is intended to provide a standard for low power transmitting devices in or around the human body. It should be noted that BAN is not designed exclusively for medical applications. Some entertainment functions for consumer electronics are also supported by BAN related technologies. From Figure A.2, we can see that different types of BAN devices can have very different bandwidth and power consumption requirements. Next, we look at some typical specifications in Table A.1.

Bluetooth and ZigBee are both types of IEEE 802.15 'Wireless Personal Area Networks' that operate in the 2.4-GHz unlicensed frequency band. Although they are similar in many ways, we outline some of their differences in Table A.2.

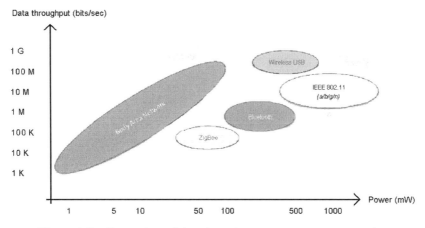

Figure A.2   Comparison of data throughput versus power consumption

Table A.1   Typical BAN specifications

Range	Up to 2 m, can be extended to 5 m
Maximum device per network	< 100
Network density per m^2	< 4
Power consumption (nominal)	1 mW / Mbps
Latency	10 ms
Device start up time	< 100 μs
Network set up time	< 1 s

Table A.2   Bluetooth vs. Zigbee

	Bluetooth	Zigbee
Standard	IEEE 802.15.1	IEEE 802.15.4
Modulation	Frequency Hopping Spread Spectrum (FHSS)	Direct Sequence Spread Spectrum (DSSS)
Network throughput	< 1 Mbps	< 250 Mbps
Typical time to establish connection	3 s	30 ms
Protocol stack size	250 KB	28 KB
Battery charging mode	Intended for frequent recharging	High capacity, low usage for prolonged use

Table A.3 Local Area Networks

		Data Rate (Mbps)	Indoor Range (m)	Outdoor Range (m)
IEEE 802.1	Bridging			
IEEE 802.3	Ethernet	100		
IEEE 802.11a	5 GHz carrier	54	35	120
IEEE 802.11b	2.4 GHz carrier	11	38	140
IEEE 802.11g	2.4 GHz carrier	54	38	140
IEEE 802.11n	2.45 GHz carrier	150	70	250

Local Area Network (LAN)

Variants of IEEE 802 (http://grouper.ieee.org/groups/802/), covering both wired and wireless networks, the collection includes those listed in Table A.3.

In essence, IEEE 802.3 covers wired networks in areas of a few hundred metres. The size of a LAN varies from a few computers in a single office to hundreds or even thousands of devices across buildings in close proximity. Connection can be made both with cables and radio waves. A LAN can also be connected to the Internet or a Metropolitan Area Network (MAN) of broader geographical coverage.

Metropolitan Area Network (MAN)

A MAN connects multiple geographically nearby LANs together as a larger overall network. It is often used for providing a 'Last-Mile' Broadband Solution within a locality. Currently, the IEEE 802.16 (http://www.ieee802.org/16/) and 802.20 (http://www.ieee802.org/20/) standards, as compared in Table A.4, are adopted for wireless metropolitan-area networks. There is also ETSI HiperMAN (http://portal.etsi.org/bran/Summary.asp), the corresponding standard of the European Telecommunications Institute, developed in conjunction with the respective IEEE groups and the WiMAX forum (Yang, 2007). Although the IEEE and ETSI standards may be similar, the European version addresses spectrum access below 10 GHz whereas IEEE's fixed WiMAX specify carriers in the 10–66 GHz range.

Table A.4 Metropolitan Area Networks

	IEEE 802.16 Wireless Broadband	IEEE 802.20 Mobile Broadband Wireless Access
Latest version (as of January 2010)	IEEE 802.16j-2009	IEEE 802.20-2008
Maximum data rate	100 Mbps mobile/1 Gbps fixed	80 Mbps mobile
Spectrum	2–11 GHz mobile/10–66 GHz fixed	< 3.5 GHz mobile
Channel bandwidth	1.25–20 MHz	5, 10, and 20 MHz
Mobility	Supports adjunct mobility services	Full mobility at up to 250 km/h
PHY	Extensions to previous 802.16a	New PHY optimized for packet data and adaptive antennas

Wide Area Network (WAN)

A WAN connects different types of network together to cover a large area. In addition to the difference in coverage area, a WAN and MAN also differs in the sense that a MAN is usually a dedicated network exclusively used by an organization or entity whereas a WAN is a shared network that is typically leased through a service provider. There is currently no standard for WANs and they vary in terms of implementation, either through a leased line or a shared line by either 'circuit switching' or 'packet switching'.

In a circuit-switched network, network resources are static. A connection is established from the sender to receiver before the start of the transfer, thereby forming a 'circuit'. The resources are dedicated to the circuit during the entire transfer and all the data follows the same path. In a packet-switched network, the data is fragmented into a number of packets each containing part of the data; each packet can take a different route across the network to the destination where the packets are reassembled into the original data at the receiver.

The Internet is perhaps the most dominant WAN across the world. Scalability is one key feature of WAN as it can be expanded to cover more areas and more devices by different means of network expansion that consists of both shared and dedicated leased lines. The key features of shared and leased lines are summarized as follows:

Shared packet switched network (generally suitable for smaller enterprise, e.g. regional hospitals):

Flexibility: Coverage expansion and access bandwidths can be easily changed without service disruption. Good for temporary site coverage as service is usually provided on a fixed-term basis.

Cost effectiveness: On-demand allocation of network resources thereby optimizing utilization efficiency with less wastage.

Consolidation: Network access is provided using the same access service at each site that consolidates circuits for cost saving.

Leased line (best solution for large enterprises, e.g. national system covering different states and provinces):

High data throughput: connection speeds in the magnitude of Gbps with quality of service (QoS) assurance.

Uncontended: Exclusive/dedicated connection with guaranteed bandwidth, predictable and stable performance.

Management: Network management for resource allocation and performance monitoring.

References

Farley, T. (2007), The Cell-Phone Revolution. *American Heritage of Invention & Technology*, New York: American Heritage, 22(3):8–19.
Huray, P. G. (2009), *Maxwell's Equations*, Wiley-IEEE Press, ISBN: 9780470542767.
Yang, X. (2007), *WiMAX/MobileFi: Advanced Research and Technology*, Auerbach Publications, ISBN: 142004351X.

Index

Telemedicine Technologies: Information Technologies in Medicine and Telehealth   Bernard Fong, A.C.M. Fong, and C.K. Li
© 2011 John Wiley & Sons, Ltd

Printed and bound by CPI Group (UK) Ltd, Croydon, CR0 4YY